an electronic companion to
genetics™ workbook

an electronic companion to
genetics™ workbook

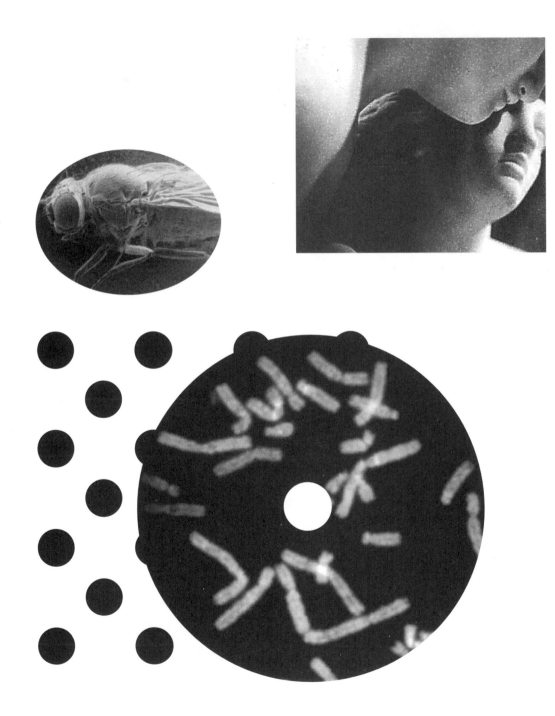

an electronic companion to
genetics™ workbook

Philip Anderson
Department of Genetics
University of Wisconsin, Madison

Barry Ganetzky
Department of Genetics
University of Wisconsin, Madison

COGITO

Cogito Learning Media, Inc.
New York   San Francisco

The authors would like to gratefully acknowledge Dr. Anita Klein, Department of Biochemistry and Molecular Biology, University of New Hampshire, for important and timely contributions to Topics 11, 12, and 13.

Cover Photo Credits: upper left, Tanya Wolff; upper right, Sky Bergman; bottom, Ulli Weier.

© **1997, Cogito Learning Media, Inc.**   ISBN: 1-888902-39-6

contents

an electronic companion to
genetics™ workbook

Mendelian Inheritance

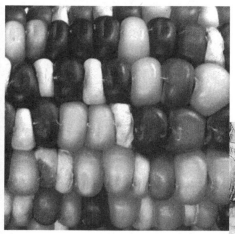

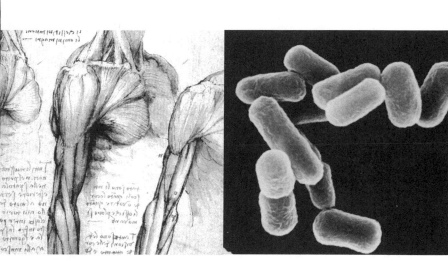

Summary

Gregor Mendel began the science of genetics with his landmark studies concerning the inheritance of physical traits of the garden pea *Pisum sativum*. Mendel studied inheritance of seven easily distinguished traits, such as tall versus short plants and round versus wrinkled seeds. His work, done over a period of ten years in a monastery in the present-day Czech Republic, represents the foundations of modern genetics. By carefully designing crosses and insightfully analyzing numerical results, Mendel formulated two principles of heredity that have become known as **"Mendel's laws."**

Mendel's laws pertain to the behavior of **genes** and **alleles** during sexual reproduction. Genes determine hereditary traits. Alleles represent alternative forms of specific genes. All nonreproductive cells of an organism contain two alleles for each gene. Such cells (and individuals) are said to be **diploid**. If the two alleles of an individual are identical, the individual is said to be **homozygous.** If the two alleles are not identical, the individual is said to be **heterozygous.** Cells involved in sexual reproduction (**gametes**) are special. Such **haploid** cells contain only one allele of each gene. Union of two haploid gametes at fertilization restores the diploid content of a one-celled embryo. Mendel's laws, in fact, describe the special behavior of genes and alleles during formation of haploid gametes.

Mendel's first law, the Principle of Segregation, states that during formation of gametes the two alleles of a gene segregate (separate) from each other such that half of the gametes carry one allele and half carry the other allele. When a plant heterozygous for alleles of a single gene (a **monohybrid**) is allowed to self-fertilize, random union of gametes yields a 1:2:1 ratio of **genotypes** among the offspring. Because alleles can be either **dominant** or **recessive**, the 1:2:1 ratio of genotypes is usually manifested as a 3:1 ratio of **phenotypes**.

Mendel's second law, the Principle of Independent Assortment, states that during formation of gametes the segregation of alleles of one gene occurs independently of the segregation of alleles of other genes. A plant heterozygous for alleles of two different genes (a **dihybrid**) produces four different genotypes of haploid gametes. Random union of such gametes yields diploid offspring of nine different genotypes and, because of dominance, four different phenotypes (in a ratio of 9:3:3:1). The **Punnett square** is a useful tool for predicting the genotypes and phenotypes resulting from crosses in which the parents differ by one or two genes.

As you will discover during your genetics course, many principles unknown to Mendel influence the rules of inheritance. Mendel's experiments were "simple" in that each trait he studied was affected by two alleles of a single gene. Mendel's first and second laws, together with the concepts of dominance and recessiveness, fully explain all of the patterns of inheritance that Mendel observed. Inheritance, however, is not always so simple, and several complicating factors must be routinely considered. Such modifications of Mendelian principles are often termed "Extensions to Mendel." **Incomplete dominance** refers to situations in which the phenotype of heterozygotes is intermediate in appearance between the phenotypes of the two respective homozygotes. **Codominance** refers to situations in which heterozygotes express a novel phenotype representing the combined phenotypes of *both* of its constituent alleles. Because of the distinctive appearance of heterozygotes in situations of incomplete dominance and codominance, the 1:2:1 ratio of genotypes in a monohybrid cross is directly apparent as a 1:2:1 ratio of phenotypes. Only two alternative alleles of each gene were known to Mendel. More frequently, a population of individuals collectively contains many different alleles of each gene (**multiple alleles**). Any *single* individual contains at most two alleles, but collectively the population can exhibit a large amount of allelic variation. In Mendel's experiments, each plant or seed trait was influenced by alleles of a single gene. However, two or more different genes can affect the same phenotype. In such cases, the phenotypic expression of one gene can mask or modify the expression of another gene or even result in novel phenotypes. Such examples of **gene interactions** often cause deviations from classic Menedelian phenotypic ratios.

Understanding genetics requires mastery of some simple rules of probability. Assortment of alleles during formation of gametes and union of gametes to form zygotes are chance events. The genotypes and phenotypes of offspring often cannot be predicted with certainty. Rather, a geneticist can only assign probabilities to certain outcomes. The probability of event A, termed $p(A)$, is the proportion of the occurrence of this event in a large number of trials. This fraction, which will always have a value between 0 and 1, can also be thought of as the probability that event A will occur in a single trial. If $p(A)$ is the probability that event A will occur, then the probability that event A will *not* occur is $1-p(A)$.

If the combined probabilities of two or more events are being calculated, care must be taken concerning whether to apply the Addition (Sum) Rule or the Multiplication (Product) Rule. If events A and B are such that the occurrence of event A precludes the occurrence of event B and, likewise, occurrence of event B precludes event A, then events A and B are said to be "mutually exclusive." For two mutually exclusive events, the probability of occurrence of *either* event A *or* event B is the sum of the individual probabilities: $p(A \text{ or } B) = p(A) + p(B)$. This rule, called the Addition (or Sum) Rule, applies to any number of mutually exclusive events (three, four, five, etc.). If, on the other hand, the occurrence of event A has no influence on whether event B will occur (and vice versa), then events A and B are said to be "independent." For two independent events, the probability that *both* events will occur is the product of their individual probabilities: $p(A \text{ and } B) = p(A) \times p(B)$. The joint occurrence of events A and B can be either simultaneous or sequential in time. This rule, called the Multiplication (or Product) Rule, can apply to any number of independent events. The Addition and Multiplication Rules are especially useful for genetics because they can be used to calculate the expected proportions of genotypes and phenotypes more quickly than by using a Punnett square.

Self-Testing Questions

1. Use the words on the left to complete the sentence on the right.

allele(s)
diploid
gamete(s)
gene
haploid

A _____ organism contains two _____ of each _____ in all cells except its _____, which are _____ and contain only one.

Answer

A <u>diploid</u> organism contains two <u>alleles</u> of each <u>gene</u> in all cells except its <u>gametes</u>, which are <u>haploid</u> and contain only one.

2. Use the words on the left to complete the sentence on the right.

dominant
homozygous
non-identical
heterozygous
identical

All non-reproductive cells contain two alleles of each gene. When those alleles are _____, the cell is said to be _____. When those alleles are _____, the cell is said to be _____, in which case the cell's (or organism's) phenotype is dictated by the _____ allele.

Answer

All non-reproductive cells contain two alleles of each gene. When those alleles are <u>identical</u>, the cell is said to be <u>homozygous</u>. When those alleles are <u>non-identical</u>, the cell is said to be <u>heterozygous</u>, in which case the cell's (or organism's) phenotype is dictated by the <u>dominant</u> allele.

3. How many different types of gametes are made by a parent of genotype *Bb*?

Answer 2

4. Is the following statement true or false? If allele *A* is dominant to allele *a*, genotypes *AA* and *Aa* exhibit the same phenotype.

Answer True

5. From the cross *Bb* × *bb*, where *B* is dominant to *b*, how many different phenotypes are produced among the offspring and in what proportions are they expected?

Answer 2, in a ratio of 1:1

6. From the cross *Bb* × *Bb*, where *B* is dominant to *b*, how many different genotypes are produced among the offspring and in what proportions are they expected?

Answer 3, in a ratio of 1:2:1

7. The dominant allele *A* causes seeds of maize to be colored, whereas the recessive allele *a* causes seeds to be colorless. From the cross *Aa* × *Aa*, what fraction of the seeds are colorless?

Answer ¼

8. Cystic fibrosis in humans is caused by a recessive allele. Two normal parents have a child with cystic fibrosis. What is the probability that this couple's next child will be normal?

➤ *Solution*

Answer: ¾

Let's abbreviate the cystic fibrosis allele as *cf* and the normal allele as *CF*. Because cystic fibrosis is a recessive trait, the affected child is genotype *cf/cf*. Because the parents are unaffected, they must both be carriers of the cystic fibrosis trait (genotypes *CF/cf*). From the cross *CF/cf* × *CF/cf*, offspring are expected to be a 1:2:1 ratio of genotypes *CF/CF*:*CF/cf*:*cf/cf*. Only children of genotype *cf/cf* (¼ of the total) are affected with cystic fibrosis. Thus, the probability that the couple's next child will be normal is 1 − ¼ = ¾.

9. A homozygous garden pea having round seeds (genotype *RR*) is crossed with a homozygous plant having wrinkled seeds (genotype *rr*). The F1 are then crossed with *RR* plants. What are the resulting phenotypes and in what proportions are they expected?

   Answer All round

10. A homozygous garden pea having yellow seeds (genotype *YY*) is crossed with a homozygous plant having green seeds (genotype *yy*). The F1 are allowed to self-fertilize, producing F2 seeds. What are the phenotypes of the F2 seeds and in what proportions are they expected?

➤ *Solution*

Answer: ¾ yellow; ¼ green

The described cross is identical to one of Mendel's monohybrid crosses. When pea plants of genotype *YY* are crossed to those of genotype *yy*, the F1 are all of genotype *Yy*. When these F1 are allowed to self-fertilize (*Yy* × *Yy*), the F2 seeds are a 1:2:1 ratio of genotypes *YY*:*Yy*:*yy*. Genotypes *YY* and *Yy* are yellow, and genotype *yy* is green. Thus, ¾ of the F2 are yellow seeds and ¼ are green seeds.

11. The tyrosinase gene, *TYR*, is necessary for production of the skin pigment melanin in humans. A recessive allele, *tyr*, of this gene causes the most frequent form of albinism.

Homozygotes (*tyr/tyr*) do not produce melanin, while heterozygotes (*TYR/tyr*) are unaffected. A normal man and woman have an albino child. What is the probability that a second child born to this couple will be albino?

Answer ¼

12. The dominant allele *R* causes seeds of maize to be colored, whereas the recessive allele *r* causes seeds to be colorless. From the cross *Rr* × *Rr*, what fraction of the colored seeds are heterozygous?

➤ **Solution**

Answer: ⅔

The described cross is equivalent to Mendel's monohybrid cross. Offspring from the cross *Rr* × *Rr* are a 1:2:1 ratio of genotypes *RR*:*Rr*:*rr*. Only genotypes *RR* and *Rr* are colored. Considering only the colored seeds, ⅔ are heterozygous (*Rr*) and ⅓ are homozygous (*RR*).

13. With rare exception, human eye color is inherited as if a dominant allele (*B*) causes eyes to be brown and the corresponding recessive allele (*b*) causes eyes to be blue. A blue-eyed man marries a brown-eyed woman whose mother had blue eyes. What proportion of their children are expected to have blue eyes?

Answer ½

14. If allele *S* is completely dominant to *s*, a plant of genotype *SS* exhibits the same phenotype as one of genotype *Ss*. In order to distinguish a plant of genotype *SS* from one of genotype *Ss*, you could cross it with a plant of genotype:
(a) *SS*
(b) *Ss*
(c) *ss*
(d) Either **(a)** or **(b)**
(e) Either **(a)** or **(c)**
(f) Either **(b)** or **(c)**

Answer **(f)**

15. Mice of genotype *yy* have gray coat color, while those of genotype *Yy* have yellow coats. *YY* homozygotes die *in utero* as small embryos well before birth.

(a) What living offspring are expected from a mating between a yellow and a gray mouse?

(b) What living offspring are expected from a mating between two yellow mice?

➤ *Solutions*

(a) *Answer:* ½ gray; ½ yellow

The cross *Yy* (yellow) × *yy* (gray) is equivalent to Mendel's mono-hybrid testcross. The *Yy* heterozygous parent produces two types of gametes (*Y* and *y*), while the *yy* homozygous parent produces only one (*y*). The offspring are expected to be a 1:1 ratio of *Yy* (yellow) and *yy* (gray).

(b) *Answer:* ⅓ gray; ⅔ yellow

The cross *Yy* (yellow) × *Yy* (yellow) is equivalent to Mendel's monohybrid cross. Offspring are expected to be a 1:2:1 ratio of genotypes *YY* : *Yy* : *yy*. In this case, however, the *YY* homozygotes die as small embryos and are absent among living offspring. Thus, living offspring are expected to be a 2:1 ratio of *Yy* (yellow) and *yy* (gray).

16. Use words on the left to complete the sentence on the right.

alleles
assortment
dominance
gametes
segregation
zygotes

The Principle of Independent Assortment (Mendel's Second Law) states that during formation of _____, the _____ of one pair of _____ occurs independently of the segregation of other pairs.

Answer

The Principle of Independent Assortment (Mendel's Second Law) states that during formation of <u>gametes</u>, the <u>segregation</u> of one pair of <u>alleles</u> occurs independently of the segregation of other pairs.

17. How many different types of gametes are produced by an individual of genotype *AaBBCc*? Assume that the three genes assort independently.

Answer 4

18. From the cross *AaBb × AaBb* (where *A* is dominant to *a*, *B* is dominant to *b*, and the two genes assort independently), how many different genotypes are present among the offspring?

Answer 9

19. Consider the cross *AaBbCc × AaBbCc*, where *A*, *B*, and *C* are dominant to *a*, *b*, and *c*, respectively, and all three genes assort independently. What proportion of the offspring will be heterozygous for all three genes?

Answer ⅛

20. How many different types of gametes are produced by an individual of genotype *AaBbCCddEeFFGg*? Assume that all 7 genes assort independently.

Answer 16

21. Consider the cross *AaBbCCDdEE × AabbCcDdee*, where all genes assort independently. How many different genotypes are present among the progeny?

Answer 36

22. With rare exception, human eye color is inherited as if a dominant allele (*B*) causes brown eyes and the corresponding recessive (*b*) causes blue eyes. Ability to taste the chemical phenyl-thio-carbamide (PTC) is inherited with a dominant allele (*A*) causing ability to taste and the corresponding recessive (*a*) causing inability to taste. The *B,b* and *A,a* genes assort independently.

    (a) What is the probability that a couple, both of whom are genotype *AaBb*, will have a child who is a blue-eyed taster?

(b) What is the probability that a couple, both of whom are genotype *AaBb*, will have a child who is a blue-eyed non-taster?

➤ **Solutions**

(a) *Answer:* $3/16$

The genotype of a blue-eyed taster is either *AAbb* or *Aabb*. From the cross *AaBb* × *AaBb*, what proportion of the offspring are either of these genotypes? Because the *A,a* and *B,b* genes assort independently, we can consider the probabilities of each gene separately. From the cross *Aa* × *Aa*, $3/4$ of the offspring are either *AA* or *Aa*; from the cross *Bb* × *Bb*, $1/4$ are *bb*. Because inheritance of *A,a* occurs independently of *B,b*, the Product Rule applies when calculating the combined probability of both genotypes. Thus $3/4 × 1/4 = 3/16$ of the offspring are either genotype *AAbb* or *Aabb*.

(b) *Answer:* $1/16$

The genotype of a blue-eyed non-taster is *aabb*. From the cross *AaBb* × *AaBb*, what proportion of the offspring are *aabb*? Because the *A,a* and *B,b* genes assort independently, we can consider the probabilities of each gene separately. From the cross *Aa* × *Aa*, $1/4$ of the offspring are *aa*; from the cross *Bb* × *Bb*, $1/4$ are *bb*. Because inheritance of *A,a* occurs independently of *B,b*, the Product Rule applies when calculating the combined probability of both genotypes. Thus $1/4 × 1/4 = 1/16$ of the offspring are genotype *aabb*.

23. As summarized below, deafness in humans can be caused by recessive alleles of either of two independently assorting genes.

Genotype	Phenotype
D- E-	Normal hearing
dd--	Deafness due to defects of the ear cochlea
--ee	Deafness due to defects of the auditory nerve

Two deaf parents have a child with normal hearing. Indicate from the list below all possible genotypes for the parents. Check all that apply.

(a) *DDEe*

(b) *DDee*

(c) *DdEE*

(d) *DdEe*
(e) *ddEe*
(f) *ddee*

Answer (b), (e)

24. Consider the cross *AaBb* × *AaBb*, where *A* is dominant to *a*, *B* is dominant to *b*, and the two genes assort independently. Among offspring having the phenotype "*AB*," how many different genotypes are present?

Answer 4

25. Consider the cross *AaBbCc* × *AaBbCc*, where *A*, *B*, and *C* are dominant to *a*, *b*, and *c*, respectively, and all three genes assort independently. What proportion of the offspring will exhibit the phenotype "*ABC*"?

Answer $^{27}/_{64}$

26. Dominant alleles of three independently-assorting maize genes (*A*, *C*, and *R*) are necessary for the development of colored seeds. Seeds of genotype *A- C- R-* are colored; all others are colorless. A plant of genotype *AaCcRr* is crossed with a plant of genotype *AaCcrr*. What proportion of the seeds will be heterozygous for all three genes?

➤ *Solution*

Answer: $^{1}/_{8}$

Offspring heterozygous for all three genes are genotype *AaCcRr*. From the cross *AaCcRr* × *AaCcrr*, what proportion of the seeds are genotype *AaCcRr*? Because all three genes assort independently, we can consider the probabilities of each gene separately. From the cross *Aa* × *Aa*, ½ of the offspring are heterozygous. From the cross *Cc* × *Cc*, ½ of the offspring are heterozygous. From the cross *Rr* × *rr*, ½ of the offspring are heterozygous. Inheritance of each gene occurs independently of the others. Thus, the Product Rule applies when calculating the combined probability of all three genotypes. Thus, ½ × ½ × ½ = ⅛ of the offspring are genotype *AaCcRr*.

27. Retinitis pigmentosa, a form of blindness, can be caused by either a recessive allele of the A,a gene or a dominant allele of the B,b gene. The two genes (A,a and B,b) assort independently. The following genotypes and phenotypes describe the inheritance of retinitis pigmentosa.

Genotype	Phenotype
A- bb	Normal vision
aa--	Retinitis pigmentosa
--B-	Retinitis pigmentosa

A woman with retinitis pigmentosa and whose parents have normal vision marries a man of genotype *AaBb*. What proportion of their children are expected to have retinitis pigmentosa?

Answer $\frac{3}{4}$

28. In Drosophila, the recessive allele *pr* causes purple eye color, while the corresponding dominant allele, *pr*$^+$, causes red eyes. The recessive allele *e* causes ebony body color, while the corresponding dominant allele, *e*$^+$, causes brown bodies. The *purple* and *ebony* genes assort independently. *pr*$^+$*pr*$^+$ *ee* females are mated with *prpr e*$^+$*e*$^+$ males, and the F1 are mated with each other.
   (a) What fraction of the F2 will have red eyes and brown bodies?
   (b) What fraction of the F2 will have red eyes and ebony bodies?

➤ **Solutions**
   (a) *Answer:* $\frac{9}{16}$

   F1 resulting from the cross *pr*$^+$*pr*$^+$ *ee* × *prpr e*$^+$*e*$^+$ are heterozygous for both genes (genotypes *pr*$^+$*pr e*$^+$*e*). When these F1 are crossed with each other (*pr*$^+$*pr e*$^+$*e* × *pr*$^+$*pr e*$^+$*e*), what fraction of the F2 have red eyes and brown bodies? Because inheritance of purple versus red eyes occurs independently of ebony versus brown body, we can consider inheritance of each gene separately. Red eyes result from genotypes *pr*$^+$*pr*$^+$ and *pr*$^+$*pr*. From the cross *pr*$^+$*pr* × *pr*$^+$*pr*, $\frac{3}{4}$ of the offspring have red eyes. Brown bodies result from genotypes *e*$^+$*e*$^+$ and *e*$^+$*e*. From the cross *e*$^+$*e* × *e*$^+$*e*, $\frac{3}{4}$ of the offspring have brown bodies. Because the *purple* and *ebony* genes assort independently, the Product Rule applies when calculating the combined probability of having both red eyes and brown bodies. Thus $\frac{3}{4} \times \frac{3}{4} = \frac{9}{16}$ of the offspring have red eyes and brown bodies.

(b) *Answer:* $\frac{3}{16}$

> From the cross $pr^{+}pr\ e^{+}e \times pr^{+}pr\ e^{+}e$, what fraction of the F2 have red eyes and ebony bodies? Red eyes result from genotypes $pr^{+}pr^{+}$ and $pr^{+}pr$. From the cross $pr^{+}pr \times pr^{+}pr$, $\frac{3}{4}$ of the offspring have red eyes. Ebony bodies result only from genotype *ee*. From the cross $e^{+}e \times e^{+}e$, $\frac{1}{4}$ of the offspring have ebony bodies. Because the *purple* and *ebony* genes assort independently, the Product Rule applies when calculating the combined probability of having both red eyes and ebony bodies. Thus $\frac{3}{4} \times \frac{1}{4} = \frac{3}{16}$ of the offspring have red eyes and ebony bodies.

29. A plant of genotype *AaBbCc* is allowed to self-fertilize. Assume that *A*, *B*, and *C* are dominant to *a*, *b*, and *c*, respectively, and that all three genes assort independently. What proportion of the offspring will have the same phenotype as the parent plant?

Answer $\frac{27}{64}$

30. In Jimsonweed, purple flowers (*P*) is dominant to white flowers (*p*), and spiny pods (*S*) is dominant to smooth pods (*s*). The *P,p* and *S,s* genes assort independently. From the cross *Pp Ss* × *pp Ss* what proportion of the progeny will have white flowers and spiny pods?

➤ **Solution**

Answer: $\frac{3}{8}$

Because the *P,p* and *S,s* genes assort independently, we can consider their inheritance separately. White flowers result from genotype *pp*. From the cross *Pp* × *pp*, $\frac{1}{2}$ of the offspring have white flowers. Spiny pods result either from genotype *SS* or *Ss*. From the cross *Ss* × *Ss*, $\frac{3}{4}$ of the offspring have spiny pods. Because inheritance of flower color occurs independently of pod shape, the Product Rule applies when calculating the combined probabilities of inheriting both white flowers and spiny pods. Thus, $\frac{1}{2} \times \frac{3}{4} = \frac{3}{8}$ of the offspring have white flowers and spiny pods.

31. In guinea pigs, rough coat (*R*) is dominant to smooth coat (*r*) and black coat (*B*) is dominant to white coat (*b*). The two genes assort independently. A homozygous rough white guinea pig is crossed with a homozygous smooth black one.

The F1 are crossed among themselves. What proportion of the F2 have smooth white coats?

Answer $\frac{1}{16}$

32. Dominant alleles of three different maize genes (A, C, and R) are necessary for the development of colored seeds. Seeds of genotype A- C- R- are colored; all others are colorless. A plant of unknown genotype is crossed with three colorless plants of known genotype. The following results are observed:

When crossed with	Result
aaccRR	50% colored, 50% colorless seeds
aaCCrr	25% colored, 75% colorless seeds
AAccrr	50% colored, 50% colorless seeds

What is the unknown plant's genotype?

Answer *AaCCRr*

33. Is the following statement true or false? "If the probability of event A occurring is $p(A)$, then the probability of event A not occurring is $1-p(A)$."

Answer True

34. Each die of a pair of dice contains 6 sides. If a single die is rolled once, what is the probability that any number other than 3 will face upwards?

Answer $\frac{5}{6}$

35. A standard deck of playing cards contains 52 cards (4 suits, 13 cards each). Without looking, you withdraw a single card from a shuffled deck. What is the probability that the selected card is either a three of clubs, a seven of spades, or any heart?

➤ ***Solution***
Answer: $\frac{15}{52}$

The probability that the chosen card is a three of clubs is $\frac{1}{52}$. The probability that it is a seven of spades is also $\frac{1}{52}$. The probability that it is any heart is $^{13}/_{52}$. Because these three outcomes are mutually exclusive events, the Sum Rule applies to calculating the probability that either event will occur. Thus the probability that the chosen card will be either a three of clubs, a seven of spades, or any heart is $^{1}/_{52}$ + $^{1}/_{52}$ + $^{13}/_{52}$ = $^{15}/_{52}$.

---

36. A large number of colored marbles are in a bowl. $^{1}/_{3}$ of the marbles are red, $^{1}/_{3}$ are white, and $^{1}/_{3}$ are blue. Without looking, you withdraw 3 marbles from the bowl. What is the probability that none of the 3 will be red?

Answer $^{8}/_{27}$

37. Phenylketonuria (PKU) in humans is caused by a recessive allele. Two normal parents have a child affected with PKU. If this couple has two additional children, what is the probability that both of them will be affected by PKU?

➤ *Solution*

Answer: $^{1}/_{16}$

Because PKU is inherited as a recessive trait, both of the unaffected parents must be heterozygous for the PKU allele. From a cross of two heterozygotes, $^{1}/_{4}$ of the children are expected to be PKU homozygotes. Each of the 2 additional children represents an independent trial, and in each trial the probability of a PKU homozygote is $^{1}/_{4}$. Thus, the probability that *both* additional children will be PKU homozygotes is $^{1}/_{4} \times ^{1}/_{4} = ^{1}/_{16}$.

---

38. α-Thallasemia, a form of anemia, is inherited in humans as a recessive trait. Two normal parents have a child affected with α-thallasemia. If this couple has 3 additional children, what is the probability that none of these 3 children will be affected by α-thallasemia?

Answer $^{27}/_{64}$

39. Galactosemia, a metabolic disorder in humans, is inherited as a recessive trait. A man and his wife are both heterozygous for the galactosemia trait. They plan to have 5 children. What is the probability that at least one of their children will be affected by galactosemia?

➤ **Solution**

Answer: $^{781}/_{1024}$

From a cross between 2 heterozygotes, ¼ of the children are expect-ed to be galactosemia homozygotes. What is the probability of having 5 unaffected children? Each child represents an independent trial, and in each trial the probability of the child being unaffected is ¾. The probability that all 5 children will be unaffected is $(¾)^5 = {}^{243}/_{1024}$. If we call this probability *x*, then 1–*x* is the probability that at least 1 child is affected by galactosemia. The value 1–*x* includes all occurrences where either 1, 2, 3, 4, or 5 children are galactosemia homozygotes. Thus the probability that at least 1 child will be affected with galac-tosemia is $1 - ({}^{243}/_{1024}) = {}^{781}/_{1024}$.

---

40. You throw 10 coins in the air simultaneously. What is the probability that the 10 coins will land such that there are 6 of one type (either heads or tails) and 4 of the other type?

Answer $^{105}/_{256}$

41. Albinism in humans is inherited as a recessive trait. A man and his wife are both heterozygous for the albinism trait. They plan to have 4 children. What is the probability that they will have exactly 2 albino and 2 non-albino children (in any order of birth)?

➤ **Solution**

Answer: $^{27}/_{128}$

The probability that each child will be an albino is ¼, and the proba-bility that it will not be an albino is ¾. There are many different orders of birth yielding exactly 2 albino and 2 non-albino children. The prob-ability of each of these could be separately calculated and then summed together, but an easier way of calculating the combined probabilities is by applying the binomial formula. The binomial formu-la states that the probability of obtaining a combination of *k* events of type A and *n*–*k* events of type B in a total of *n* trials is:

$$\frac{n!}{k!(n-k)!}\, p^k q^{n-k}$$

where
p = probability of event type A, and
q = probability of event type B = 1–*p*.

In this example:

$n = 4$

k = the number of albino children = 2

$n-k$ = the number of non-albino children = 2

p = the probability of having an albino child = ¼

q = the probability of having a non-albino child = ¾

Thus, the calculated probability of having exactly 2 albino and 2 non-albino children is

$$\frac{4!}{(2!)(2!)}(¼)^2(¾)^2 = \frac{27}{128}$$

42. ABO blood types in humans are determined by a single gene having three alleles, I^A, I^B, and i.

Genotype	Blood Type
I^AI^A and I^Ai	A
I^BI^B and I^Bi	B
I^AI^B	AB
ii	O

(a) A mother with blood type A has a child with blood type O. What is the mother's genotype?

(b) A mother with blood type B has a child with blood type O. What are all possible genotypes of the father?

(c) A woman of blood type AB marries a man of blood type AB. What blood types are possible among their children?

(d) A woman of blood type A marries a man of blood type B. Their first child is blood type O. What is the probability that their second child will have a blood type other than O?

Answer

(a) I^Ai

(b) I^Ai, I^Bi, or ii

(c) Types A, B, and AB

(d) ¾

43. As summarized in the table below, two genes affect feather color of chickens. Allele C (dominant) causes feathers to be colored, while allele c (recessive) causes feathers to be white. Allele I (dominant) prevents the formation of color (even when C is present), while allele i (recessive) has no effect on color formation. The C,c and I,i genes assort independently.

Genotype	Phenotype
C- ii	Colored
cc--	White
-- I-	White

Hens of genotype *CC II* are mated with roosters of genotype *cc ii*. What are the phenotypes of F1 chicks and in what proportions are they expected?

Answer All chicks are white

44. Retinitis pigmentosa (RP), a form of blindness, can be caused by alleles of either of two genes that assort independently. Consider two alleles of each gene, designated here as alleles A^1 and A^2 of gene "A," and alleles B^1 and B^2 of gene "B." The following genotypes and phenotypes describe the inheritance of RP:

Genotype	Phenotype
A^1- B^1B^1	Normal vision
A^2A^2 --	Retinitis pigmentosa
-- B^2-	Retinitis pigmentosa

Which of the following statements is true? Check all that apply.
(a) Allele A^1 is dominant to allele A^2
(b) Allele A^2 is dominant to allele A^1
(c) Alleles A^1 and A^2 are codominant
(d) Allele B^1 is dominant to allele B^2
(e) Allele B^2 is dominant to allele B^1
(f) Alleles B^1 and B^2 are codominant

Answer (a), (e)

45. As summarized in the table below, coat color in cocker spaniels is governed by alleles of two genes. The *A,a* and *B,b* genes assort independently.

Genotype	Phenotype
A- B-	Black coat
aa B-	Liver-colored coat
A- bb	Red coat
aa bb	Lemon-colored coat

A black cocker spaniel male is mated to a lemon-colored female and produces a lemon-colored pup. If the black male is mated to a female of its own genotype, what proportion of the offspring are expected to have liver-colored coats?

➤ *Solution*

Answer: $3/16$

To solve this problem we must first deduce the genotype of the black male. When the black male was mated with a lemon-colored female (genotype *aabb*), a lemon-colored pup (genotype *aabb*) resulted. The black male must have contributed the recessive alleles *a* and *b* to the lemon-colored pup. This is only possible if the genotype of the black male is *AaBb*.

Having established the genotype of the black male, the initial question can now be rephrased, "From the cross *AaBb* × *AaBb*, what proportion of the offspring are genotype *aa B-*?" Because the *A,a* and *B,b* genes segregate independently, we can consider them separately in calculating the probability of *aa B-* offspring. From the cross *Aa* × *Aa*, $1/4$ of the offspring are genotype *aa*. From the cross *Bb* × *Bb*, $3/4$ of the offspring are genotype *B-*. Because inheritance of *A,a* and *B,b* occurs independently, the Product Rule applies when calculating the combined probability of inheriting both *aa* and *B-*. Thus $1/4 × 3/4 = 3/16$ of the pups are expected to be liver-colored.

46. As summarized in the table below, alleles of two genes determine coat color and the extent of coat coloration in cattle. The dominant allele *B* determines black coat color, whereas the recessive allele *b* determines red. The extent of coloration on the body is determined by the codominant alleles *R* and *r*. *RR* homozygotes have solid coloration, *Rr* heterozygotes have colored coats spotted with white ("roan"), and *rr* homozygotes have solid white coats (regardless of the *B,b* genotype). The *B,b* and *R,r* genes assort independently.

Genotype	Phenotype
B- RR	black solid
B- Rr	black roan
bb RR	red solid
bb Rr	red roan
-- rr	solid white

Cows of genotype *bbRr* are mated with bulls of genotype *BbRr*. What proportion of the calves will have solid coats (either black, red, or white)?

➤ *Solution*

Answer: $1/2$

Solid coats result from either genotypes *RR* (colored) or *rr* (white). In the case of *RR* homozygotes, the color of the solid coat is determined by alleles of the *B,b* gene, but the extent of coloration (solid versus roan versus white) is determined only by alleles of the *R,r* gene. Thus, the initial question can be rephrased, "From the cross *Rr* × *Rr*, what proportion of calves are genotype *RR* or *rr*?" This cross is equivalent to Mendel's monohybrid cross. ¼ of the offspring are expected to be genotype *RR*, and ¼ are expected to be *rr*. Thus, ¼ + ¼ = ½ of the calves are expected to have solid coats.

---

47. The dominant allele *B* determines black coat color in cattle, whereas the recessive allele *b* determines red. The extent of coloration on the body is determined by the codominant alleles *R* and *r*. *RR* homozygotes have solid coloration, *Rr* heterozygotes have colored coats spotted with white ("roan"), and *rr* homozygotes have solid white coats (regardless of the *B,b* genotype). The dominant allele *P* causes animals to be hornless ("polled"), whereas the recessive allele *p* causes horns to grow. All three genes assort independently. To summarize:

Genotype	Phenotype	Genotype	Phenotype	Genotype	Phenotype
B-	Black	*RR*	solid color	*P-*	Polled
bb	Red	*Rr*	roan	*pp*	Horned
		rr	white		

Red roan polled bulls (genotype *bbRrPp*) are mated with black roan horned cows (genotype *BbRrpp*). What proportion of the calves will be red solid polled?

Answer ¹⁄₁₆

The Chromosomal Basis of Heredity

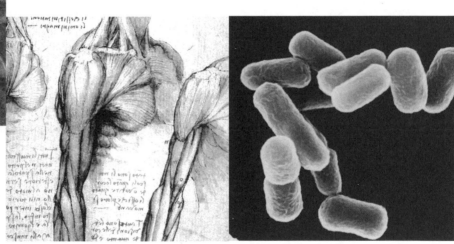

Summary

Mitosis consists of the cellular processes by which chromosomes are distributed to daughter cells during cell division. Mitosis, together with other important events of the cell division cycle such as DNA replication, insures that hereditary information is faithfully transmitted from one cell to its daughter cells. "Mitosis" strictly refers only to **M phase**, that portion of the cell division cycle in which chromosomes condense, the nuclear envelope breaks down, and chromosomes move into newly forming nuclei. The entire cycle of cell division is referred to as the "mitotic" cell cycle. Mitotic cell division is accomplished by the orderly execution of several sequential processes:

1. DNA replication produces two identical copies of each chromosome;
2. replicated chromosomes separate from each other; and
3. separated chromosomes move to opposite poles of a dividing cell.

Accurate segregation of chromosomes requires assembly and function of a complex mechanical machine, the **mitotic spindle**. The mitotic spindle is responsible for insuring faithful chromosome segregation. **Centromeres**, the sites at which spindle **microtubules** attach to chromosomes, are instrumental in this process. Centromeres and **kinetochores** are the cellular structures that are distributed so precisely during mitosis. The chromosomes, in a sense, just "go along for the ride."

The goal of mitotic cell division is to faithfully and accurately transmit genetic information from one cell to its descendent cells. The two daughter cells produced by mitosis are identical to the parent that produced them. In principle, all somatic cells of the body are exact copies of the one-celled embryo because of the precision and reproducibility of mitosis.

Meiosis, on the other hand, does not preserve intact a cell's genetic information. Rather, meiosis reduces by half the amount of genetic material contained in a cell. Meiosis is a specialized form of cell division that occurs only during formation of **gametes**, the **haploid** reproductive cells that all sexually-reproducing species have. Meiosis, for example, occurs during formation of sperm and ova in mammals, during formation of pollen and ovules in plants, and during formation of spores in fungi. No matter what the species, when two haploid gametes fuse during fertilization, chromosome content is restored to its normal **diploid** state.

Chromosome number is reduced by half during meiosis because a single round of DNA replication is followed by two rounds of cell division (**meiosis I** and **meiosis II**). The division of chromosomes by meiosis is an orderly process:

1. DNA replication produces two identical copies of each chromosome.
2. **Homologous** pairs of chromosomes align with each other.
3. Paired homologues attach to a spindle whose structure is very similar to that found in mitosis.
4. The two members of a pair move to opposite poles of the dividing cell.
5. Each replicated chromosome attaches to a second spindle and divides in a manner very similar to that of mitosis.

Following completion of meiosis, gametes contain one (and only one) set of chromosomes. When two gametes fuse together at fertilization, the resulting zygote contains two full sets of chromosomes.

Meiosis serves two important biological roles. First, as described above, meiosis reduces the number of chromosomes from the diploid to the haploid number. This is an important part of maintaining the constancy of genetic information from one generation to the next. Chromosome number is reduced by half at meiosis and restored to diploid levels at fertilization. Thus, species maintain their genetic endowment over long periods of time. Second, meiosis "reshuffles the genetic deck." The manner in which any single chromosome pair aligns at meiosis occurs independently of the manner in which any other chromosome pair aligns. Whereas a diploid cell entering meiosis contains two copies of each chromosome, a gamete formed by meiosis contains only one copy of each chromosome. But the specific *combination* of chromosomes in each gamete is a random process. Thus, meiosis reassorts the genes on one chromosome with all of the other genes on other chromosomes.

The events of meiosis are the cellular underpinnings of Mendel's laws of heredity. Mendel's first law, the Principle of Segregation, follows from the fact that homologous chromosomes pair during meiosis I and move to opposite poles of the dividing cell. Each gamete receives one (and only one) chromosome of the pair. Mendel's second law, the Principle of Independent Assortment, follows from the fact that pairing and separation of any single chromosome pair occurs independently of the pairing and segregation of other chromosome pairs present in the same cell. Different genes, therefore, assort independently because different chromosomes pair and segregate independently.

Self-Testing Questions

1. Indicate with a check in the appropriate column whether the following structures or processes occur during mitosis, during meiosis, or during both mitosis and meiosis.

	Mitosis	Meiosis	Both Mitosis and Meiosis
bivalents	___	___	___
centrioles	___	___	___
centromeres	___	___	___
chiasmata	___	___	___
chromatids	___	___	___
cytokinesis	___	___	___
kinetochores	___	___	___
microtubules	___	___	___
nondisjunction	___	___	___
reductional division	___	___	___
spindles	___	___	___
synapsis	___	___	___
synaptonemal complexes	___	___	___
telophase	___	___	___

Answer

	Mitosis	Meiosis	Both Mitosis and Meiosis
bivalents	___	X	___
centrioles	___	___	X
centromeres	___	___	X
chiasmata	___	X	___
chromatids	___	___	X
cytokinesis	___	___	X
kinetochores	___	___	X
microtubules	___	___	X
nondisjunction	___	___	X
reductional division	___	X	___
spindles	___	___	X
synapsis	___	X	___
synaptonemal complexes	___	X	___
telophase	___	___	X

2. Sperm of the laboratory rat, *Rattus norvegicus*, contain 22 chromosomes.

   (a) How many chromatids are contained in rat somatic cells at prophase of mitosis?

   (b) How many chromosomes are contained in rat meiotic cells at metaphase of meiosis II?

➤ *Solutions*

   (a) *Answer:* 88

     Sperm are gametes and contain the haploid number of chromosomes. If sperm contain 22 chromosomes, then somatic cells, which are diploid, contain 44 chromosomes. At prophase of mitosis, each chromosome is composed of two sister chromatids. Sister chromatids are the products of DNA replication, which occurs during interphase, prior to mitosis. At prophase of mitosis, sister chromatids are connected to each other at their centromeres. Thus, a *R. norvegicus* somatic cell contains 88 chromatids at prophase of mitosis.

   (b) *Answer:* 22

     As discussed above, *R. norvegicus* somatic cells contain 44 chromosomes. When such cells enter meiosis, homologous chromosomes pair with each other, thereby forming 22 bivalents. Meiosis I is the reductional division of meiosis, meaning that the number of chromosomes is reduced by half. Homologues separate from each other at anaphase of meiosis I and move to opposite poles of the dividing cell. At the completion of meiosis I, therefore, each cell (a secondary meiocyte) contains 22 chromosomes, each of which is composed of two sister chromatids. These 22 chromosomes line up at the metaphase plate during metaphase of meiosis II. Thus, a *R. norvegicus* cell at metaphase of meiosis II contains 22 chromosomes.

3. The following diagram depicts a dividing cell at anaphase.

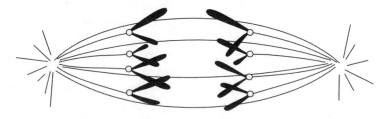

(a) Assume that this diagram comes from an animal whose diploid number of chromosomes is 8. Which of the following mitotic or meiotic phases could this diagram depict? Check all that are possible.

___ anaphase of mitosis
___ anaphase of meiosis I
___ anaphase of meiosis II

(b) Assume that this diagram comes from an animal whose diploid number of chromosomes is 4. Which of the following mitotic or meiotic phases could this diagram depict? Check all that are possible.

___ anaphase of mitosis
___ anaphase of meiosis I
___ anaphase of meiosis II

➤ *Solutions*

(a) *Answer:* anaphase of meiosis II

During the anaphase depicted, 4 chromosomes are moving toward each pole of the dividing cell. Because the diploid number of chromosomes is 8, this cannot be a cell in anaphase of mitosis. If it were, 8 (not 4) chromosomes would be moving toward each pole. The cell must be in meiosis. But in which meiotic division is it? During anaphase I, homologous chromosomes, each of which is composed of two sister chromatids, separate from each other. Because the chromosomes pictured above are not composed of sister chromatids, this cannot be a cell in anaphase of meiosis I. Rather, the cell must be in anaphase II. During anaphase II, sister chromatids separate from each other, which is what is depicted above.

(b) *Answer:* anaphase of mitosis

During the anaphase depicted, 4 chromosomes are moving toward each pole of the dividing cell. Because the diploid number of chromosomes is 4 in this case, this can only be a cell in anaphase of mitosis. Neither meiosis I or II is consistent with the above diagram. For animals in which $2n = 4$, cells in anaphase I of meiosis would contain two replicated chromosomes moving toward each pole. The chromosomes depicted above are not composed of sister chromatids, so this cannot be meiosis I. Cells in anaphase II of meiosis would contain only two chromosomes moving toward each pole. Thus, this cell cannot be a meiotic cell.

4. Put a check in the appropriate column to indicate in which phase of the mitotic cell cycle each of the following events or processes occurs.

	Interphase	Prophase	Metaphase	Anaphase	Telophase
chromosome condensation	___	___	___	___	___
DNA replication	___	___	___	___	___
attachment of chromosomes to the spindle	___	___	___	___	___
movement of chromosomes toward the spindle poles	___	___	___	___	___
nuclear envelope breakdown	___	___	___	___	___
alignment of chromosomes at the center of the cell	___	___	___	___	___
centromere separation	___	___	___	___	___
reformation of the nuclear envelope	___	___	___	___	___

Answer

	Interphase	Prophase	Metaphase	Anaphase	Telophase
chromosome condensation	___	_X_	___	___	___
DNA replication	_X_	___	___	___	___
attachment of chromosomes to the spindle	___	_X_	___	___	___
movement of chromosomes toward the spindle poles	___	___	___	_X_	___
nuclear envelope breakdown	___	_X_	___	___	___
alignment of chromosomes at the center of the cell	___	___	_X_	___	___
centromere separation	___	___	___	_X_	___
reformation of the nuclear envelope	___	___	___	___	_X_

5. At what stage of mitosis is the following cell?

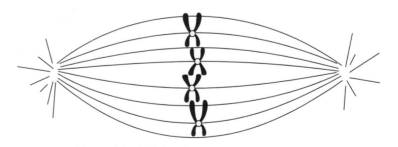

Answer metaphase

6. Complete the following sentences:
    (a) The end of a chromosome is called the _____.

(b) The site on a chromosome to which spindle microtubules attach is called the _____.

(c) The process in which homologous chromosomes pair with each other during meiosis is called _____.

(d) The products of DNA replication, which remain attached to each other at the centromere until anaphase, are called _____.

(e) During mitosis, separation of centromeres marks the onset of _____.

(f) Meiosis I is called the _____ division of meiosis.

(g) Meiosis II is called the _____ division of meiosis.

Answer
(a) telomere
(b) centromere or kinetochore
(c) synapsis
(d) chromatids or sister chromatids
(e) anaphase
(f) reductional
(g) equational

7. The following diagram depicts a dividing cell at anaphase of meiosis I.

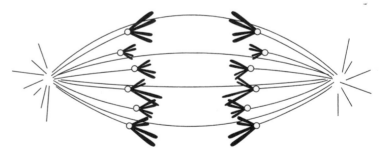

What is the diploid number of chromosomes for this species?

Answer 12

8. The diploid number of chromosomes in the African zebra, *Equus burchelli*, is 44.

(a) How many telomeres are contained in zebra somatic cells during the G1 phase of interphase?

(b) How many telomeres are contained in zebra somatic cells during the G2 phase of interphase?

(c) How many telomeres are contained in zebra somatic cells during metaphase of mitosis?

(d) How many telomeres are contained in zebra meiotic cells following completion of meiosis I?

(e) How many telomeres are contained in zebra meiotic cells following completion of meiosis II?

Answer

(a) 88

(b) 176

(c) 176

(d) 88

(e) 44

9. The haploid number of chromosomes in the snapdragon, *Antirrhinum majus,* is 8.

(a) How many chromosomes are contained in snapdragon meiotic cells following the completion of meiosis I but before the onset of meiosis II?

(b) How many chromatids are contained in the same cells?

Answer

(a) 8

(b) 16

10. Dwarfism in tomatoes is caused by a recessive allele, designated *d*. A dwarf plant is crossed with a normal plant. All of the offspring are normal. These normal plants are crossed with a dwarf plant. What proportion of the offspring will be dwarf?

Answer $\frac{1}{2}$

11. A plant of genotype *AaBb* is allowed to self-fertilize. The *A,a* and *B,b* genes assort independently.

(a) What proportion of the offspring are genotype *AaBb*?

(b) What proportion of the offspring are genotype *aabb*?

(c) What proportion of the offspring are homozygous for both the *A,a* and *B,b* genes?

(d) What proportion of the offspring are homozygous for one gene (either *A,a* or *B,b*) and heterozygous for the other gene?

Answer
(a) ¼
(b) ¹⁄₁₆
(c) ¼
(d) ½

12. Both Manx anury (taillessness, designated *M*) and polydactylia (extra digits, designated *Pd*) in cats are caused by dominant mutations. The *M* and *Pd* genes assort independently. A heterozygous Manx female (genotype *M/+*) is crossed with a heterozygous polydactyl male (genotype *Pd/+*).

(a) What proportion of the offspring are expected to be tailless and have normal digits?

(b) What proportion of the offspring are expected to have extra digits and normal tails?

(c) What proportion of the offspring are expected to be tailless and have extra digits?

Answer
(a) ¼
(b) ¼
(c) ¼

13. The diploid number of chromosomes in the Mexican axolotyl is 28. When a male axolotyl undergoes meiosis, what proportion of his sperm contain the same set of chromosomes (centromeres) as he inherited from his father?

Answer $(½)^{14}$

14. The diploid number of chromosomes in guppies, *Poecilia reticulata*, is 46. Two guppies that are heterozygous for genetic markers on every chromosome (e.g., genotype *Aa Bb Cc Dd...Tt Uu Vv Ww*, where all genes segregate independently) are crossed.

(a) With respect to these markers, how many genetically distinct types of gametes are produced as a result of independent assortment during meiosis?

(b) How many possible genotypes of offspring are produced by the above cross?

Answer
(a) 2^{23}
(b) 3^{23}

15. As shown in the table below, coat color in rabbits is governed by four alleles of a single gene.

Allele	Coat Color
C	full color
c^{chm}	chinchilla
c^h	Himalayan
c	albino (white)

C is dominant to c^{chm}, c^h, and c
c^{chm} is dominant to c^h and c
c^h is dominant to c

(a) A female with full color is mated with a chinchilla male. Two of the offspring have full color, 1 is chinchilla, and 1 is albino. What are the genotypes of the parents?
(b) A female of genotype Cc is mated with a male of genotype $c^{chm}c^h$. What proportion of the offspring are Himalayan?

Answer
(a) The female is Cc, and the male is $c^{chm}c$.
(b) ¼

16. As shown in the table below, fruit shape in summer squash, *Cucurbita pepo*, is determined by alleles of two genes (A,a and B,b) that assort independently.

Genotype	Shape of fruit
A-B-	disc
A-bb or aaB-	sphere
aabb	elongate

(a) A plant of genotype *AaBb* is crossed to a plant of genotype *Aabb*. What proportion of the offspring have elongate fruit?
(b) A plant of genotype *AaBb* is crossed to a plant of genotype *aabb*. What proportion of the offspring have sphere fruit?
(c) A plant with sphere fruit is crossed to a plant with elongate fruit. Half of the offspring have sphere fruit, and half have elongate fruit. What are the possible genotypes of the parent with sphere fruit?

(d) A plant with disc fruit is crossed to a plant with elongate fruit. One fourth of the offspring have elongate fruit, ¼ have disc fruit, and ½ have sphere fruit. What is the genotype of the parent with disc fruit?

(e) A plant of unknown genotype is crossed with a plant of genotype *AAbb*. Half of the offspring have disc fruit and half have sphere fruit. The same unknown plant is crossed with a plant of genotype *aaBB*. All of the offspring have disc fruit. What is the genotype of the unknown plant, and what shape are its fruit?

(f) Two plants with sphere fruit are crossed. All of the offspring have disc fruit. What are the genotypes of the two parent plants?

➤ *Solutions*

(a) *Answer:* ⅛

Elongate fruit are produced only by plants of genotype *aabb*. Thus, this question can be rephrased, "What proportion of the offspring have genotype *aabb*?" The genotypes of the parents are given in the problem (*AaBb* and *Aabb*). The problem also states that the *A,a* and *B,b* genes assort independently. Thus, if we calculate the proportion that are *aa*, and multiply by the proportion that are *bb*, the product will be the proportion that are both *aa* and *bb*. Considering only the *A,a* gene, the cross is *Aa* × *Aa*. This is equivalent to Mendel's monohybrid cross; ¼ of the offspring are *aa*. Considering only the *B,b* gene, the cross is *Bb* × *bb*. This is equivalent to Mendel's monohybrid testcross; ½ of the offspring are *bb*. Thus (¼)(½) = ⅛ of the offspring are genotype *aa bb* and have elongate fruit.

(b) *Answer:* ½

Sphere fruit are produced by plants either of genotype *A-bb* or of genotype *aaB-*. Both of these genotypes must be considered when calculating the proportion of offspring having sphere fruit. These two genotypes are mutually exclusive outcomes of the cross. Thus, the Sum Rule applies when calculating the combined probability. That is, the proportion of offspring that are genotype *A- bb* is added to the proportion that is genotype *aaB-*. The sum represents the proportion that is either *A-bb* or *aaB-*.

Calculating the proportion of offspring that is genotype *A-bb* is done in a manner similar to that described in part (a). The two genes are considered separately, and each independent proportion is determined. Then, the independent proportions are multiplied together. Considering only the *A,a* gene, the cross is *Aa* × *aa*. One-half of the offspring are genotype *A-*. Considering only the *B,b* gene, the cross is *Bb* × *bb*. Again, ½ of the offspring are genotype *bb*. Thus, (½)(½) = ¼ of the offspring are genotype *A-bb*. This number (¼) is one of the two we need to calculate the proportion of sphere fruit.

The second number we need is the proportion of offspring that are genotype *aaB-*. This is calculated in a similar manner. From the cross *AaBb* × *aabb*, ½ of the offspring are genotype *aa*, and ½ are genotype *B-*. Thus, (½)(½) = ¼ are genotype *aaB-*.

We now must add together the individual probabilities of being genotype *A-bb* (¼) and *aaB-* (¼). Thus, ½ of the offspring are either of these genotypes and have sphere fruit.

(c) *Answer: Aabb* or *aaBb*

This question asks you to determine the genotype of the plant with sphere fruit. There are only four possibilities (*AAbb*, *Aabb*, *aaBB*, and *aaBb*), and we must examine the information provided to determine which of these genotypes is consistent with the outcome of the cross described. The cross is sphere × elongate. Although we don't know for certain the genotype of the parent plant with sphere fruit, the parent with elongate fruit can only be genotype *aabb*. The cross yields ½ plants with sphere fruit and ½ plants with elongate fruit. Which of the four possible genotypes for the sphere-fruited parent is consistent with this result? Genotypes *AAbb* or *aaBB* are not consistent with the observations. Either of these genotypes would produce only offspring with sphere fruit when crossed with *aabb*. This leaves only *Aabb* and *aaBb* as possible genotypes for the sphere-fruited plant. The 1:1 ratio suggests segregation of a single heterozygous gene, similar to Mendel's monohybrid testcross. Both *Aabb* and *aaBb* are heterozygous for a single gene. When either of these two genotypes is crossed with *aabb*, the offspring are half sphere and half elongate. Thus, both of these genotypes are consistent with the observations. Without additional information we cannot distinguish between them. The parent with sphere fruit could be either of these two genotypes.

(d) *Answer: AaBb*

This question asks you to determine the genotype of the parent with disc fruit. There are only four possibilities (*AABB*, *AABb*, *AaBB*, and *AaBb*). We must evaluate the information provided and determine which of these four genotypes is consistent with the outcome of the cross. The cross is disc × elongate. Although we don't know the genotype of the parent with disc fruit, the parent with elongate fruit can only be genotype *aabb*. The cross yields offspring of three different phenotypes (¼ elongate, ¼ disc, ½ sphere). Which of the four possible genotypes for the disc-fruited parent is consistent with this result? Only one! When genotypes *AABB*, *AABb*, or *AaBB* are crossed with *aabb*, the offspring are either all disc-fruited or half disc- and half sphere-fruited plants. Neither of these genotypes yields plants having disc, sphere, and elongate fruit. This leaves *AaBb*. What are the offspring from the cross *AaBb* × *aabb*? One-fourth are genotype *AaBb* (disc), ¼ are *Aabb* (sphere), ¼ are *aaBb* (sphere), and ¼ are *aabb* (elongate). Thus, the parent plant with disc fruit must be genotype *AaBb*.

(e) *Answer: AABb,* disc fruit

In this question, we are not even provided the phenotype of the parent plant of unknown genotype. We must deduce its phenotype by figuring out its genotype. When the unknown plant is crossed with *AAbb*, half of the offspring have disc fruit and half have sphere fruit. How could the offspring with sphere fruit be obtained? All gametes of the *AAbb* parent contain an *A* allele. Thus, with regard to the *A,a* gene the offspring are genotype *A-*. This implies that the offspring with sphere fruit must be genotype *A-bb*. This occurs only if the sphere-fruited offspring receives a *b* allele from the unknown parent. We deduce, therefore, that the unknown parent contains a *b* allele as part of its genotype. But, is it homozygous *bb*, or heterozygous *Bb*? If it were homozygous *bb*, then *all* offspring would be plants with sphere fruit. Because half are sphere-fruited and half are disc-fruited, the unknown parent must be a *Bb* heterozygote. At this point, we have deduced the genotype of the unknown parent with regard to the *B,b* gene. What about the *A,a* gene? The cross we just considered provides no information concerning the *A,a* gene.

The second cross described in the question, however, allows us to deduce the unknown plant's *A,a* genotype in a manner similar to that described above. When the unknown is crossed with *aaBB*, all of the offspring have disc fruit. Plants with disc fruit are obtained only if the unknown parent contributes *A* alleles. Is the unknown homozygous *AA* or heterozygous *Aa*? If it were heterozygous *Aa*, then half the offspring would have sphere fruit (genotype *aaB-*), and half would have disc fruit (genotype *AaB-*). If it were homozygous *AA*, then all of the offspring would have disc fruit (genotype *AaB-*). This latter possibility is exactly what is observed in the cross. Thus, the unknown parent must be an *AA* homozygote.

By combining information deduced from both of the crosses described, we have now deduced the unknown parent's genotype. It is genotype *AABb* and bears disc-shaped fruit.

(f) *Answer: AAbb* and *aaBB*

Two plants with sphere fruit are crossed, and all of the offspring have disc fruit. There are four possible genotypes for each of the sphere-fruited parents (*AAbb, Aabb, aaBB,* and *aaBb*). It would be a daunting task to consider all possible combinations of genotypes (there are 16) and determine which of them produces only off-spring with disc fruit. Rather, this question is more easily solved "by inspection."

We know from the information provided as part of the question that a plant with sphere fruit is either homozygous *aa* or *bb* (but not both). We also know that plants with disc fruit are genotype *A-B-*. Thus, if the sphere-fruited parent were genotype *aaB-*, it con-tributes only *a* alleles, and the *A* allele of disc-fruited offspring must come from the other parent. Similarly, if the sphere-fruited parent were genotype *A-bb*, it contributes only *b* alleles, and the *B* allele of disc-fruited offspring must come from the other parent. At this point we have narrowed down the possible genotypes of the parents. One is genotype *aaB-*, and the other is genotype *A-bb*.

Is the *aaB-* parent homozygous *BB* or heterozygous *Bb*? If it were heterozygous *Bb*, when crossed with *A-bb* some of its off-spring would have either sphere or elongate fruit. Such offspring are not observed. Thus, one parent must be genotype *aaBB*. By similar reasoning, we can deduce that the *A-bb* parent must be genotype *AAbb*. When sphere-fruited plants of genotype *aaBB* are crossed with sphere-fruited plants of genotype *AAbb*, all offspring are genotype *AaBb* and bear disc fruit. This is the only combination of genotypes that yields such results.

17. Coat color in cattle is governed by alleles of a single gene. A dominant allele, *B*, causes black color, while the corre-sponding recessive, *b*, causes red color. Codominant alleles of a second, independently-assorting, gene controls the extent of coloration on the body. *RR* homozygotes have uni-formly colored coats; *Rr* heterozygotes have colored coats spotted with white patches (called "roan"), and *rr* homozy-gotes have uniformly white coats (regardless of the *B,b* genotype).

(a) A black roan cow and a white bull have a red roan calf. What is the genotype of the cow?

(b) A solid black cow and a black roan bull have a red roan calf. What is the probability that a second calf born to the same parents will be solid red?

Answer

(a) The cow is genotype *Bb Rr.*

(b) ⅛

18. Leaf shape in tomatoes is controlled by alleles of two genes that assort independently. A dominant allele, *cu*, causes leaves to be curly, while the corresponding recessive, *cu⁺*, causes leaves to be straight. A recessive allele, *hl*, causes leaves to be hairless, while the corresponding dominant, *hl⁺*, causes leaves to be hairy.

(a) A curly heterozygous plant is allowed to self-fertilize. What proportion of the offspring are expected to have straight leaves?

(b) A plant heterozygous for both curly and hairy is allowed to self-fertilize. What proportion of the offspring are expected to have curly hairy leaves?

(c) A curly hairless plant is crossed with a straight hairy plant. Half of the offspring are curly hairy and half are straight hairy. What is the genotype of the curly hairless parent?

(d) A straight hairy plant is crossed with a curly hairless plant. Half of the offspring are curly hairy and half are straight hairy. What is the genotype of the straight hairy parent?

Answer

(a) ¼

(b) 9⁄16

(c) *cucu⁺ hlhl*

(d) *cu⁺cu⁺ hl⁺hl⁺*

19. The shape and color of radishes is controlled by alleles of two genes, designated *L,l* and *R,r*. *LL* homozygous radishes are long, *Ll* heterozygotes are oval, and *ll* homozygotes are round. *RR* homozygotes are red, *Rr* heterozygotes are purple, and *rr* homozygotes are white.

(a) A long, red radish plant is crossed with a round, white plant. What are the phenotypes of the F1, and in what proportions are they expected?

(b) An F1 plant from part (a) (above) is allowed to self-fertilize. What phenotypes are expected among the offspring, and in what proportions are they expected?

(c) An F1 radish of unknown genotype is crossed with a round, red radish. Half of the offspring are plants with oval, purple radishes, and half are plants with round, purple radishes. What is the genotype and phenotype of the unknown plant?

(d) A radish of unknown genotype is crossed with an oval, purple radish. The offspring are a 1:1:1:1 ratio of plants with oval red, round red, oval purple, and round purple radishes. What is the genotype and phenotype of the unknown plant?

Answer

(a) All F1 plants have oval purple radishes.

(b) The offspring are $\frac{1}{16}$ long red, $\frac{1}{8}$ long purple, $\frac{1}{16}$ long white, $\frac{1}{8}$ oval red, $\frac{1}{4}$ oval purple, $\frac{1}{8}$ oval white, $\frac{1}{16}$ round red, $\frac{1}{8}$ round purple, and $\frac{1}{16}$ round white.

(c) The unknown plant is genotype *Ll rr* and produces oval white radishes.

(d) The unknown plant is genotype *ll RR* and produces round red radishes.

20. Formation of seed color in maize is governed by alleles of three independently-assorting genes, designated *A,a, C,c,* and *R,r. A* is dominant to *a, C* is dominant to *c,* and *R* is dominant to *r.* Seeds of genotype *A-C-R-* are colored; all others are colorless. A plant of genotype *AaCcRr* is allowed to self-fertilize.

(a) What proportion of the pollen contain alleles *A, C,* and *R*?

(b) How many different genotypes are present among the offspring?

(c) What proportion of the offspring are colorless?

(d) How many different genotypes are colorless?

Answer

(a) $\frac{1}{8}$

(b) 27

(c) $\frac{37}{64}$

(d) 19

topic 3
Sex Chromosomes

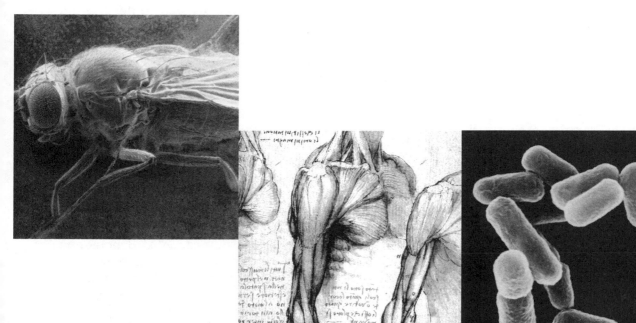

Summary

Males and females of most species are strikingly different. Development of an embryo either as a male or as a female is governed by key developmental regulatory genes that are located on a special pair of chromosomes, the **sex chromosomes**. Sex chromosomes are unique in that the two members of the pair can be very different in size and in appearance when viewed with a microscope. Male mammals, for example, contain one X and one Y chromosome, while females contain two X chromosomes. The mammalian X and Y chromosomes are distinctly different from each other in size, in morphology, and, as discussed below, in the genes that each of them carries. Understanding that the sexual fate of an individual is associated with inheritance of a distinct pair of chromosomes was instrumental in the 1900s in proving that genes are located on chromosomes.

For all chromosomes except the sex chromosomes, the two members of each pair of chromosomes are morphologically identical to each other in both males and females. Such chromosomes are termed **autosomal chromosomes** or **autosomes.** Humans, for example, contain twenty-two pairs of autosomes and two sex chromosomes (fourty-six total). The **karyotype** of an individual is a description of the complete set of chromosomes that it contains. Male mammals, for example, are karyotype AA XY, where "X" and "Y" denote the sex chromosomes and "A" denotes one full set of autosomes. Such males are said to be **heterogametic**, because their gametes contain either an X or a Y chromosome. Female mammals (karyotype AA XX) are **homogametic**, because all of their gametes contain an X chromosome. In some species— birds, for example—females are heterogametic and males are homogametic. Sex chromosomes in such species are usually termed "Z" and "W." Females are karyotype AA ZW; males are karyotype AA ZZ.

In almost all species, there are approximately equal numbers of males and females. This constancy of sex ratio is caused by the behavior of sex chromosomes during meiosis. The X and Y chromosomes pair as homologues and segregate from each other during meiosis I. Thus, half of male sperm contain one full set of autosomes plus an X chromosome, and half contain one full set of autosomes plus a Y chromosome. Female eggs contain one full set of autosomes plus an X chromosome. When X-containing sperm fertilize an egg, the offspring (karyotype AA XX) are female. When Y-containing sperm fertilize an egg, the offspring (karyotype AA XY) are male. Equal numbers of X- and Y-containing sperm yield equal numbers of males and females.

In both mammals and Drosophila, embryos of karyotype AA XX develop as females, and those of karyotype AA XY develop as males. The manner in which chromosome content specifies sexual fate, however, is different in these two organisms. Sex of mammals is determined by a **dominant-Y** mechanism. Key developmental regulatory genes are located on the mammalian Y chromosome. If such genes are present in an embryo, it develops as male. If such genes are absent, it develops as female. Thus, mammals of karyotype AA XY, AA XXY, and AA XYY are males, while those of karyotype AA XX, AA XXX, and AA XO are females. Sex of Drosophila is determined by an **X-to-autosome ratio** mechanism. The key developmental regulatory genes of Drosophila are located on both the X chromosome and autosomes. The *number* of X chromosomes relative to the *number* of sets of autosomes determines the sex of Drosophila. Embryos in which the X-to-autosome ratio is 1.0 or greater develop as females. Those in which the ratio is 0.5 or less develop as males. Thus, Drosophila of karyotype AA XX, AA XXX, and AA XXY are females, while those of karyotype AA XY, AA XYY, and AA XO are males. Drosophila with X-to-autosome ratios between 0.5 and 1.0 (e.g., AAAA XXX) develop as intersexual animals, exhibiting aspects of both males and females.

The orderly process by which chromosomes separate from each other and move to opposite poles during mitosis or meiosis is termed **disjunction.** Occasional errors of chromosome separation (**nondisjunction**) yield gametes that contain abnormal numbers of chromosomes. Individuals with abnormal numbers of sex chromosomes (eg., AA XXY, AA XYY, and AA XO) result from such nondisjunction events. When nondisjunction of sex chromosomes occurs during germline mitosis or during meiosis II, the resulting gametes contain either two X chromosomes (diplo-X), two Y chromosomes (diplo-Y), or no sex chromosomes (nullo-XY). When nondisjunction of sex chromosomes occurs during meiosis I, the resulting gametes contain either two X chromosomes, no sex chromosomes, or both an X and a Y chromosome (XY-gametes). Such exceptional gametes, produced by nondisjunction, give rise to abnormal karyotypes involving the sex chromosomes.

Genes located on the sex chromosomes have special patterns of inheritance. X chromosomes contain thousands of genes, alleles of which affect an individual's phenotype in many ways. Females contain two X chromosomes. Thus, the concepts of dominance, recessiveness, homozygosity, and heterozygosity apply to X-linked genes of females in a normal manner. Males, however, contain an X and a Y chromosome. Y chromosomes contain very few genes apart from those that are directly involved in sex determination. Thus, the concepts of homozygosity and heterozygosity do not apply in males. Males are said to be **hemizygous** for X-linked genes. Females receive X chromosomes from both their mothers and fathers. Males receive X chromosomes only from their mothers and Y chromosomes only from their fathers. An important consequence of sex chromosome inheritance is that X-linked recessive traits are evident in males much more often than in females. Hemizygous males exhibit any X-linked recessive traits that they inherit. Females, on the other hand, exhibit such traits only when they are homozygous, not when they are heterozygous. X-linked recessive traits are often said to exhibit "criss-cross" inheritance, in which affected fathers transmit the causative allele to unaffected heterozygous daughters, who in turn transmit it to affected sons. This contrasts with autosomal inheritance, in which the numbers of affected males and females are equal.

Self-Testing Questions

1. Indicate whether the following statements are true or false.
  _____ **(a)** Human X-linked genes are hemizygous in males.
  _____ **(b)** Female birds have homomorphic sex chromosomes.
  _____ **(c)** A mouse of karyotype AA XYY is male.
  _____ **(d)** *A Drosophila* of karyotype AA XXXY is male.
  _____ **(e)** In *Drosophila*, females are the heterogametic sex.
  _____ **(f)** Dominant X-linked traits are transmitted from fathers to daughters, but not from fathers to sons.
  _____ **(g)** Recessive X-linked alleles are transmitted from mothers to daughters, but not from fathers to daughters.
  _____ **(h)** Although female mammals contain two X chromosomes, only one of them is expressed in individual somatic cells.
  _____ **(i)** Nondisjunction during meiosis II generates two nullisomic and two disomic gametes.
  _____ **(j)** Nondisjunction occurring in the mitotic divisions of oogonial or spermatogonial cells (prior to meiosis) can have the same genetic consequences on resulting gametes as nondisjunction occurring during either meiosis I or meiosis II.

Answer

T   **(a)** Human X-linked genes are hemizygous in males.
F   **(b)** Female birds have homomorphic sex chromosomes.
T   **(c)** A mouse of karyotype AA XYY is male.
F   **(d)** *A Drosophila* of karyotype AA XXXY is male.
F   **(e)** In *Drosophila*, females are the heterogametic sex.
T   **(f)** Dominant X-linked traits are transmitted from fathers to daughters, but not from fathers to sons.
F   **(g)** Recessive X-linked alleles are transmitted from mothers to daughters, but not from fathers to daughters.
T   **(h)** Although female mammals contain two X chromosomes, only one of them is expressed in individual somatic cells.
F   **(i)** Nondisjunction during meiosis II generates two nullisomic and two disomic gametes.
T   **(j)** Nondisjunction occurring during the mitotic divisions of oogonial or spermatogonial cells (prior to meiosis) can have the same genetic consequences on the resulting gametes as nondisjunction occurring during either meiosis I or meiosis II.

2. Indicate the sex of the following individuals. Fill in "M" (male) or "F" (female).

	ORGANISM	
Karyotype	**Drosophila**	**Mammals**
AA XX	____	____
AA XY	____	____
AA XXX	____	____
AA XXY	____	____
AA XO	____	____

➤ *Solution*

Answer:

	ORGANISM	
Karyotype	**Drosophila**	**Mammals**
AA XX	F	F
AA XY	M	M
AA XXX	F	F
AA XXY	F	M
AA XO	M	F

In both Drosophila and mammals, the sex of an individual is determined by its chromosomal composition. The manner in which chromosomes establish sex, however, is fundamentally different in these two organisms.

In Drosophila, sex is determined by the "X-to-autosome (X:A) ratio," meaning that the number of X chromosomes relative to the number of sets of autosomal chromosomes determines whether an embryo will develop as a male or as a female. Normal females (karyotype AA XX) have an X:A ratio of 1.0 (2 X chromosomes, 2 sets of autosomes). Flies with an X:A ratio of 1.0 or greater develop as females. Thus, flies of karyotype AA XX, AA XXX, and AA XXY are females. Flies with X:A ratios considerably greater than 1.0 (e.g., AA XXX and AA XXXX) are abnormal, but they nevertheless develop as females. Normal males (karyotype AA XY) have an X:A ratio of 0.5. Flies with an X:A ratio of 0.5 or less develop as males. Thus, flies of karyotype AA XY and AA XO are males.

In mammals, sex is determined by a "dominant Y" mechanism, meaning that an embryo whose karyotype includes one or more Y chromosomes develops as a male, regardless of the number of X chromosomes it contains. A key developmental regulatory gene, *TDF/SRY,* is located on the Y chromosome and directs male development. Embryos containing Y chromosomes are male; those not containing Y chromosomes are female. Thus, mice of karyotype AA XX, AA XXX, and AA XO are female, while those of karyotype AA XY and AA XXY are male.

3. Indicate the number of Barr bodies contained in somatic cells of each of the following human karyotypes.

Karyotype	Number of Barr Bodies
AA XX	_____
AA XY	_____
AA XXX	_____
AA XXY	_____
AA XO	_____
AA XXXY	_____

Answer

Karyotype	Number of Barr Bodies
AA XX	1
AA XY	0
AA XXX	2
AA XXY	1
AA XO	0
AA XXXY	2

4. Red-green color blindness in humans is caused by recessive alleles of an X-linked gene.

(a) A colorblind woman marries a man with normal color vision. What proportion of their sons and daughters are expected to be colorblind?

(b) A normal woman whose father was colorblind marries a man with normal color vision. What proportion of their sons and daughters are expected to be colorblind?

➤ *Solutions*

(a) *Answer:* All sons will be colorblind; all daughters will have normal color vision.

Let X^g indicate an X chromosome containing a red-green color blindness allele and X^+ indicate an X chromosome containing the normal allele. The colorblind woman is genotype $X^g X^g$, and the normal man is genotype $X^+ Y$. The woman produces only a single type of ova (X^g), while the man produces 2 types of sperm (X^+ and Y). Random union of sperm and ova yields 2 genotypes among the offspring: $X^g X^+$ and $X^g Y$. Thus, all of the daughters have normal color vision (they are carriers), and all of the sons are colorblind.

(b) *Answer:* Half of the sons will be colorblind; all daughters will have normal color vision.

Let X^g indicate an X chromosome containing a red-green color blindness allele and X^+ indicate an X chromosome containing the normal allele. The woman with normal color vision inherited one of her X chromosomes from her father, who was colorblind. Thus, the woman must be heterozygous (genotype $X^g X^+$). The man with normal vision is genotype $X^+ Y$. The woman produces 2 types of ova (X^g and X^+), while the man produces 2 types of sperm (X^+ and Y). Random union of sperm and ova yield 4 genotypes among the offspring: $X^g X^+$, $X^+ X^+$, $X^g Y$, and $X^+ Y$. Thus, all of the daughters have normal color vision (½ of them are carriers), and ½ of the sons are colorblind.

5. The *barred* (striped) color pattern of chickens is inherited as a sex-linked dominant trait. Let Z^B indicate a Z chromosome containing the dominant barred allele and Z^b indicate a Z chromosome containing the corresponding recessive (non-barred) allele. From each of the following crosses, what proportion of male and female chicks are expected to have barred feathers?

(a) $Z^B Z^b$ barred rooster × $Z^b W$ nonbarred hen

(b) $Z^B Z^b$ barred rooster × $Z^B W$ barred hen

(c) $Z^b Z^b$ nonbarred rooster $\times Z^B W$ barred hen

➤ **Solutions**

Sex-linked inheritance in birds is similar to that of other organisms, but with one important difference. In birds, the sex chromosomes of females are morphologically dissimilar, while those of males are morphologically identical. Thus, female birds are said to be "heterogametic" (two different types of gametes), while males are said to be "homogametic" (one type of gamete). This contrasts with the sex chromosomes of Drosophila and mammals, for example, in which males (karyotype AA XY) are heterogametic and females (karyotype AA XX) are homogametic. To distinguish the sex chromosomes of birds (and other organisms that use a similar mechanism of sex determination) from those of other organisms, the sex chromosomes are termed "Z" and "W" (instead of "X" and "Y"). Female birds are karyotype AA ZW; males are karyotype AA ZZ.

(a) *Answer:* Half of the females and $\frac{1}{2}$ of the males are barred.

> A heterozygous barred rooster is crossed with a nonbarred hen. The rooster produces 2 types of sperm, Z^B and Z^b. The hen produces 2 types of eggs, Z^b and W. Random union of sperm and egg yields 4 different genotypes: $Z^B Z^b$, $Z^b Z^b$, $Z^B W$, and $Z^b W$. Thus, $\frac{1}{4}$ of the offspring are barred males, $\frac{1}{4}$ are nonbarred males, $\frac{1}{4}$ are barred females, and $\frac{1}{4}$ are nonbarred females.

(b) *Answer:* Half of the females are barred; all of the males are barred.

> A heterozygous barred rooster is crossed with a barred hen. The rooster produces 2 types of sperm, Z^B and Z^b. The hen produces 2 types of eggs, Z^B and W. Random union of sperm and egg yields 4 different genotypes: $Z^B Z^B$, $Z^B Z^b$, $Z^B W$, and $Z^b W$. Thus, $\frac{1}{2}$ of the offspring are barred males, $\frac{1}{4}$ are barred females, and $\frac{1}{4}$ are nonbarred females.

(c) *Answer:* All of the females are nonbarred; all of the males are barred.

> A nonbarred rooster is crossed with a barred hen. The rooster produces a single type of sperm, Z^b. The hen produces 2 types of eggs, Z^B and *W*. Random union of sperm and egg yields 2 different genotypes: $Z^B Z^b$ and $Z^b W$. Thus, $\frac{1}{2}$ of the offspring are barred males, and $\frac{1}{2}$ are nonbarred females.

6. Eye color in humans is usually inherited as if brown eyes were caused by an autosomal dominant allele and blue eyes by the corresponding recessive. (Assume that this is correct, even though the true pattern of inheritance is more complex.) Red-green color blindness is inherited as an X-linked recessive trait.

   (a) Two brown-eyed parents with normal color vision have a blue-eyed colorblind son. What are the genotypes of the parents?

   (b) A blue-eyed woman with normal color vision but whose father was color-blind marries a brown-eyed man with normal vision and whose mother had blue eyes. What proportion of their sons will be blue-eyed and colorblind?

➤ *Solutions*

In this question we must consider the simultaneous inheritance of two unlinked genes. Because they are inherited independently, we can consider them separately and, if necessary, multiply together the individual probabilities of inheritance to calculate a combined probability simultaneous inheritance. Let B and b indicate the dominant and recessive alleles causing brown and blue eyes, respectively. Let X^g and X^+ indicate X chromosomes containing a red-green color blindness allele and the normal allele, respectively.

(a) *Answer:* The mother's genotype is $Bb\ X^g X^+$, and the father's genotype is $Bb\ X^+Y$, where B indicates the autosomal dominant allele yielding brown eyes, b indicates the recessive allele yielding blue eyes, X^g indicates an X chromosome carrying the color blindness allele, X^+ indicates a normal X chromosome, and Y indicates a normal Y chromosome.

With regard to eye color, both parents have brown eyes, yet their son has blue eyes. Because blue eyes is recessive, both parents must be heterozygous (genotype Bb). With regard to color blindness, the colorblind son inherited his X chromosome only from his mother. Thus the mother, whose color vision is normal, must be heterozygous (genotype $X^g X^+$). The father, whose vision is normal, must be genotype X^+Y. Thus, the mother's genotype is $Bb\ X^g X^+$ and the father's genotype is $Bb\ X^+Y$.

(b) *Answer:* ¼ of the sons will be blue-eyed and colorblind.

With regard to eye color, the blue-eyed woman must be genotype bb and, because his mother had blue eyes, the brown-eyed man must be genotype *Bb*. With regard to color vision, the woman received one of her X chromosomes from her father, who was colorblind. Thus, she is heterozygous (genotype $X^g X^+$). The man, whose vision is normal, must be genotype X^+Y. Thus, the cross is *bb* $X^g X^+ \times Bb$ X^+Y. The question asks, "What proportion of sons will be blue-eyed and colorblind?" From the cross *bb* × *Bb*, ½ of the offspring have blue eyes. From the cross $X^g X^+ \times X^+ Y$, ½ of the sons are colorblind. Thus, ½ × ½ = ¼ of the sons will be blue-eyed and colorblind.

7. Consider an X-linked recessive lethal mutation, designated *m*, in Drosophila. Homozygous females (*m/m*) and hemizygous males (*m/Y*) die as early embryos; heterozygous females (*m/+*) are unaffected. Heterozygous females (*m/+*) are mated with normal males. Which of the following statements most accurately describes the sexes of the living offspring?
   (a) There will be twice as many males as females.
   (b) There will be twice as many females as males.
   (c) There will be equal numbers of males and females.

   Answer (b)

8. As summarized in the table below, coat color in cats is determined by two alleles (*O* and *o*) of an X-linked gene. In females, *oo* homozygotes are orange, *OO* homozygotes are black, and *Oo* heterozygotes are spotted with black and orange ("tortoiseshell"). In males, *oY* and *OY* hemizygotes are orange and black, respectively.

Females	Males
oo = orange	*oY* = orange
OO = black	*OY* = black
Oo = tortoiseshell	

   (a) A black female is mated with an orange male. What phenotypes are expected among the kittens, and in what proportions should they occur?
   (b) In a litter of eight kittens, there are two tortoiseshell females, two orange females, one black male, and three orange males. What are the phenotypes of the mother and father?

   Answer
   (a) All females are tortoiseshell; all males are black.
   (b) The mother is tortoiseshell; the father is orange.

9. An X-linked recessive allele, w, causes white eye color in Drosophila, while the normal allele, w^+, produces red eyes. An autosomal recessive allele, b, causes black body color, while the normal allele, b^+, produces brown bodies.
   (a) What phenotypes are expected among the F1 progeny of a cross between red-eyed black-bodied males and homozygous white-eyed brown-bodied females?
   (b) What phenotypes are expected among the F1 progeny of the cross $bb^+ ww^+ \times bb^+ wY$ (brown-bodied red-eyed females × brown-bodied white-eyed males)?

   Answer

   (a) All female offspring have red eyes and brown bodies; all male offspring have white eyes and brown bodies.
   (b) $1/16$ black-bodied white-eyed females; $1/16$ black-bodied white-eyed males; $1/16$ black-bodied red-eyed females; $1/16$ black-bodied red-eyed males; $3/16$ brown-bodied white-eyed females; $3/16$ brown-bodied white-eyed males; $3/16$ brown-bodied red-eyed females; $3/16$ brown-bodied red-eyed males.

10. Hemophilia A in humans (Factor VIII deficiency) is an X-linked recessive disease. A woman with hemophilia A marries an unaffected man whose father had hemophilia A. What proportion of their sons and daughters are expected to have hemophilia A?

   Answer All of their children will be unaffected.

11. Both hemophilia A (Factor VIII deficiency) and hemophilia B (Factor IX deficiency) are X-linked recessive diseases. Although the Factor VIII and Factor IX genes both reside on the X chromosome, they are different genes and are located at different positions on the X chromosome. A man with hemophilia B marries an unaffected woman. Their firstborn son has hemophilia A. What proportion of their daughters are expected to be carriers of either hemophilia A or B?

   ➤ *Solution*
   Answer: All daughters are carriers of hemophilia B. Half of the daughters are carriers of both hemophilia A and B.

Let X^A indicate an X chromosome carrying an allele of hemophilia A, X^B indicate an X chromosome carrying an allele of hemophilia B, and X^+ indicate an X chromosome carrying the normal allele of both genes. The man with hemophilia B is genotype $X^B Y$. His wife is unaffected by hemophilia, but they have a son affected with hemophilia A (genotype $X^A Y$). Because sons receive X chromosomes only from their mothers, the mother must be a carrier of hemophilia A (genotype $X^A X^+$).

Having established genotypes for the mother ($X^A X^+$) and father ($X^B Y$), we can now predict the genotypes of their daughters. Daughters receive one X chromosome from their father. Thus, all daughters are carriers of hemophilia B. Daughters also receive one X chromosome from their mothers. In this case, the daughter might receive a normal X chromosome from her mother, in which case her genotype would be $X^B X^+$, or she might receive the hemophilia A allele from her mother, in which case her genotype would be $X^B X^A$. Thus, all daughters are carriers of hemophilia B, and half of them are carriers of both hemophilia A and B. Because hemophilia A and B are caused by recessive alleles of different genes, none of the daughters is affected by hemophilia.

12. A dominant allele, *D*, of an autosomal gene causes "duplex" combs in chickens; the corresponding recessive allele, *d*, causes "simplex" combs. A dominant allele, *K*, of a Z-linked gene causes delayed feathering; the corresponding recessive allele, *k*, causes rapid feathering. A rooster having duplex comb and delayed feathering is mated with a hen having simplex comb and rapid feathering. The offspring are:

   ¼ males with duplex combs and delayed feathering
   ¼ males with duplex combs and rapid feathering
   ¼ females with duplex combs and delayed feathering
   ¼ females with duplex combs and rapid feathering

   With regard to the *K,k* gene, is the rooster homozygous, heterozygous, or hemizygous?

   Answer Heterozygous

13. Red-green color blindness is an X-linked recessive trait. Approximately 16% of western European women are carriers of red-green color blindness. What proportion of western European men are red-green colorblind?

➤ *Solution*

Answer: 8%

Men have only a single X chromosome, which they inherit from their mothers. If a mother is heterozygous for red-green color blindness, half her sons will be affected. Thus, if approximately 16% of western European women are carriers, then approximately 8% of men are red-green colorblind.

---

14. Favism, an inherited hemolytic anemia brought about by ingestion of the bean *Vicia fava* or exposure to its pollen, is caused by deficiency of the enzyme glucose-6-phosphate dehydrogenase (G6PD). G6PD is located on the X chromosome, and favism is inherited as an X-linked recessive disorder. What is the probability that a boy whose great-great grandfather (mother's mother's mother's father) had favism, will inherit the disease?

Answer $\frac{1}{8}$

---

15. Testicular feminization, *tfm,* is a rare X-linked recessive disorder caused by defects in the androgen (male hormone) receptor. Hemizygotes of genotype *tfm*/Y develop as sterile phenotypic females despite having an AA XY karyotype.

What proportion of the female offspring of a *tfm*/+ carrier mother are fertile? What proportion are sterile?

➤ *Solution*

Answer: $\frac{2}{3}$ are fertile (genotypes *tfm*/+ and +/+); $\frac{1}{3}$ are sterile (genotype *tfm*/Y)

Let X^{tfm} indicate an X chromosome carrying the *tfm* allele and X^+ indicate an X chromosome carrying the normal allele. A carrier female (genotype X^{tfm}/X^+) produces 2 types of ova (X^{tfm} and X^+). A normal male produces 2 types of sperm (X^+ and Y). Random union of ova and sperm yields 4 genotypes: $X^{tfm}X^+$, X^+X^+, $X^{tfm}Y$, and X^+Y. Genotypes $X^{tfm}X^+$ and X^+X^+ develop as fertile females (*tfm* is recessive); genotype $X^{tfm}Y$ develops as a sterile phenotypic female despite having an AA XY karyotype; genotype X^+Y develops as a normal male. Thus, $\frac{2}{3}$ of females are fertile and $\frac{1}{3}$ are sterile.

---

16. Both red/green color blindness, *g*, and testicular feminization, *tfm*, are X-linked recessive traits. In the case of testicular feminization, hemizygous individuals (genotype *tfm*/Y) develop as sterile females despite having an AA XY karyotype. A normal mother and father (normal color vision; normal karyotype) have two children. One is a daughter with testicular feminization (genotype *tfm*/Y) and normal color vision; the other is a colorblind son with a normal AA XY karyotype. If this couple has one additional child, what is the probability that it will be a colorblind daughter? (Ignore the possibility of nondisjunction and crossing over [Topic 5] in your answer.)

Answer 0

17. The coats of Ayrshire cattle are white spotted, meaning their coats contain patches of white and dark. Two alleles of a single autosomal gene control the color of the dark parts. *mahogany* homozygotes are mahogany colored in both males and females. *red* homozygotes are red colored in both males in females. *mahogany*/*red* heterozygotes are mahogany colored in males and red colored in females. Although the phenotype of *mahogany*/*red* heterozygotes is affected by the sex of the animal, the *mahogany*, *red* gene is autosomal, not sex-linked.

A *mahogany*/*red* heterozygous bull is mated with a *mahogany*/*red* heterozygous cow. What proportion of the calves are expected to be mahogany females?

➤ *Solution*

Answer: ⅛

In this question, we must consider the inheritance of both an autosomal color-determining gene (*mahogany* versus *red*) and the sex chromosomes (male versus female). From the cross *mahogany*/*red* × *mahogany*/*red*, ¼ of the offspring are *mahogany*/*mahogany*, ½ are *mahogany*/*red*, and ¼ are *red*/*red*. Within each group, ½ of the calves are male and ½ are females. Of the females, only the genotype *mahogany*/*mahogany* is colored mahogany. (*mahogany*/*red* and *red*/*red* females are colored red.) Thus ¼ × ½ = ⅛ of the calves are mahogany-colored females.

18. In magpie moths (*Abraxas*), sex is determined by a ZZ/ZW mechanism (similar to birds). Males are homogametic, and females are heterogametic. Wing color in *Abraxas* is governed by alleles of a Z-linked gene. The allele for light colored wings, *l*, is recessive to that for dark colored wings, *L*. Males with light colored wings are mated to females with dark colored wings. What phenotypes are expected among the offspring, and in what proportions should they occur?

Answer All of the F1 males have dark wings; all of the females have light wings.

19. Two parents with normal karyotypes have a daughter whose karyotype is AA XO. In which parent did nondisjunction occur to yield the AA XO daughter?
(a) the mother
(b) the father
(c) uncertain; nondisjunction could have occurred in either the mother or the father

Answer (c)

20. Two parents with normal karyotypes have a Klinefelter syndrome son (karyotype AA XXY). Nondisjunction during which of the following stages of meiosis might yield AA XXY sons?

_____ Meiosis I of the mother
_____ Meiosis II of the mother
_____ Meiosis I of the father
_____ Meiosis II of the father

➤ *Solution*
Answer:

_X_ Meiosis I of the mother
_X_ Meiosis II of the mother
_X_ Meiosis I of the father
___ Meiosis II of the father

Klinefelter males (karyotype AA XXY) have one too many sex chromosomes. Assuming that nondisjunction occurs in only one parent, there are two possible origins for the AA XXY son:

1. a diplo-X ova fertilized by a normal Y-containing sperm, or
2. a normal X-containing ova fertilized by an XY-containing sperm.

How might diplo-X ova be generated? Nondisjunction either during meiosis I or meiosis II of the mother might yield diplo-X ova. If nondisjunction occurs during meiosis I, the two X chromosomes contained in a single ovum are homologues (one copy each of the two X chromosomes contained by the mother). If nondisjunction occurs during meiosis II, the two X chromosomes are the products of DNA replication (two identical copies of one X chromosome, originally generated as sister chromatids). Although nondisjunction during meiosis II of the father can generate diplo-X sperm, this cannot contribute to formation of AA XXY sons. Diplo-X sperm yield triplo-X daughters, not XXY sons.

How might XY-containing sperm be generated? Nondisjunction during meiosis I of the father yields XY-sperm. X and Y chromosomes pair as homologues during male meiosis. Nondisjunction during meiosis I yields sperm containing both an X and Y chromosome. Nondisjunction during meiosis II of the father yields diplo-X or diplo-Y, but not XY-containing, sperm.

Thus, nondisjunction during meiosis I or II of the mother or meiosis I of the father can yield an AA XXY Klinefelter son.

---

21. Red/green color blindness is a common sex-linked recessive trait. When answering the following questions, assume that nondisjunction occurs only during meiosis.

(a) A mother and father who both have normal color vision and a normal karyotype have a colorblind son whose karyotype is AA XYY. In which parent did nondisjunction occur? Check any of the following that explain this pattern of inheritance.

    ___ Mother only
    ___ Father only
    ___ Both mother and father

(b) A mother having normal color vision and a colorblind father have a colorblind daughter whose karyotype is AA XXX. The mother and father have a normal karyotype. In which parent and at which meiotic division did nondisjunction occur? Check any of the following that explain this pattern of inheritance.

    ___ Meiosis I of the mother
    ___ Meiosis II of the mother
    ___ Meiosis I of the father
    ___ Meiosis II of the father

(c) A colorblind mother and a father having normal color vision have a son with normal color vision. The mother, father, and son have a normal karyotype. In which parent did nondisjunction occur? Check any of the following that explain this pattern of inheritance.

_____ Mother only
_____ Father only
_____ Both mother and father

(d) A colorblind mother and a father with normal color vision have a son with normal vision whose karyotype is AA XXY. The mother and father have a normal karyotype. In which parent and at which meiotic division did nondisjunction occur? Check any of the following that explain this pattern of inheritance.

_____ Meiosis I of the mother
_____ Meiosis II of the mother
_____ Meiosis I of the father
_____ Meiosis II of the father

Answer

(a) Father only
(b) Meiosis II of the mother or father
(c) Both mother and father
(d) Meiosis I of the father

22. Fabry disease is an X-linked recessive disorder caused by alleles of the gene encoding α-galactosidase A, an enzyme involved in sugar metabolism. A normal man and woman (no Fabry disease; normal karyotype) have a Klinefelter syndrome son (karyotype AA XXY) affected with Fabry disease. In which parent and in which division of meiosis did the nondisjunction occur to yield the affected son? Check any of the following that might explain the observed inheritance.

_____ Meiosis I of the mother
_____ Meiosis II of the mother
_____ Meiosis I of the father
_____ Meiosis II of the father

Answer

_____ Meiosis I of the mother
__X__ Meiosis II of the mother
_____ Meiosis I of the father
_____ Meiosis II of the father

23. Hemophilia in humans is an X-linked recessive trait. Two normal parents (normal karyotype; no hemophilia) have a hemophiliac son whose karyotype is AA XXY.

(a) What are the genotypes of the parents?

(b) In which parent did nondisjunction occur to yield the AA XXY son?

(c) In which stage of meiosis (meiosis I or II) did the nondisjunction occur?

Answer

(a) Mother = X^+X^h
    Father = X^+Y,
    where X^+ indicates a normal X chromosome, X^h indicates an X chromosome carrying the hemophilia allele, and Y indicates a normal Y chromosome.

(b) Nondisjunction occurred in the mother.

(c) Nondisjunction occurred during meiosis II of the mother.

24. Down syndrome occurs when individuals inherit 3 copies of chromosome 21 (trisomy-21). A gene encoding the liver form of the enzyme phosphofructokinase (*PFK*) is located on chromosome 21. Three codominant alleles of *PFK* are known, designated *PFK¹*, *PFK²*, and *PFK³*. A normal woman of genotype $PFK^2\ PFK^3$ marries a man of genotype PFK^1PFK^2. They have a child with Down syndrome whose genotype is $PFK^1\ PFK^2\ PFK^3$. In which parent and in which division of meiosis did the nondisjunction occur to yield the Down syndrome child? Check any of the following that might explain the observed inheritance. (Ignore the possibility of crossing over [Topic 5] in your answer.)

_____ Meiosis I of the mother
_____ Meiosis II of the mother
_____ Meiosis I of the father
_____ Meiosis II of the father

➤ **Solution**

Answer:

_X_ Meiosis I of the mother
___ Meiosis II of the mother
_X_ Meiosis I of the father
___ Meiosis II of the father

The child with Down syndrome contains an additional chromosome 21. Assuming that nondisjunction occurs in only one parent, there are two possible origins for the trisomy-21 child:

1. a diplo-21 ovum carrying both the PFK^2 and PFK^3 alleles fertilized by a normal sperm carrying the PFK^1 allele, or
2. a normal ovum carrying the PFK^3 allele fertilized with a diplo-21 sperm carrying both the PFK^1 and PFK^2 alleles.

In both cases, the two chromosomes 21 contained in an aneuploid gamete constitute one copy of two homologous chromosomes (as opposed to two identical copies of one of them). In both males and females, homologous chromosomes pair and disjoin during meiosis I. Thus, nondisjunction during meiosis I of either the mother or father can yield a Down syndrome child of genotype $PFK^1 PFK^2 PFK^3$.

Analysis
of Human Pedigrees

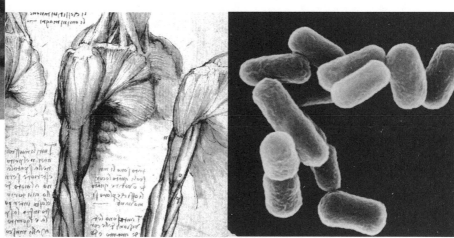

Summary

Our everyday experiences tell us that many human traits are inherited. Blue-eyed parents have blue-eyed children. Diseases often run in families. As its name implies, "human genetics" is the study of inherited variation in man. Human geneticists strive to distinguish genetic variation from non-genetic variation and to understand molecular mechanisms that underlie genetic variation. Human genetics is especially concerned with inherited disease. A great deal of human disease is directly caused by, or at least influenced by, the genes we inherit.

Genetic investigations of humans pose special challenges. The types of experimental genetics that were pioneered by Mendel and that continue today with a wide variety of model organisms are not possible in humans. Family sizes are small, making numerical analysis of affected and unaffected individuals difficult. Human "crosses" cannot be designed. Rather, human geneticists must retrospectively analyze crosses that have already occurred. The generation time of humans is very long, restricting genetic analysis to at most a few generations. It is quite understandable why most genetic principles were first elucidated with model organisms. Those principles, however, apply to humans just as precisely as they do to Mendel's garden peas.

A family tree showing the relationships between family members and indicating which individuals exhibit a certain trait is termed a **pedigree**. Human geneticists search through such pedigrees examining transmission of a trait from one generation to the next. In such information there are clues concerning whether the trait is a heritable trait and, if so, what its mode of inheritance is. "Mode of inheritance" in this context means whether the gene(s) governing a particular trait is (are) located on the autosomes or on the sex chromosomes, and which alleles of those genes are dominant or recessive. Like Sherlock Holmes solving a crime, human geneticists deduce the mechanisms of inheritance from the facts of the pedigree.

Traits caused by recessive alleles of genes located on the autosomes exhibit **autosomal recessive** inheritance. Homozygotes exhibit such traits; heterozygotes do not. The characteristic features of autosomal recessive inheritance are:

1. Affected children often have unaffected parents.
2. Two affected parents have only affected children.

3. The number of affected males and females are aproximately equal.
4. Affected individuals are more frequent among the offspring of related parents.

Examples of autosomal recessive inheritance include cystic fibrosis and albinism.

Traits caused by dominant alleles of genes located on the autosomes exhibit **autosomal dominant** inheritance. Both homozygotes and heterozygotes exhibit such traits, but for rare traits most affected individuals are heterozygotes. The characteristic features of autosomal dominant inheritance are:

1. The trait does not skip generations.
2. The number of affected males and females are approximately equal.
3. Unaffected parents have only unaffected children.
4. Affected parents (most of whom are heterozygous) have both affected and unaffected children.
5. In families where only one parent is an affected heterozygote, half the children on average are affected.

Examples of autosomal dominant inheritance include Huntington's disease and achondroplastic dwarfism.

Traits caused by recessive alleles of genes located on the X chromosome exhibit **X-linked recessive** (also called **sex-linked recessive)** inheritance. Homozygous females and hemizygous males exhibit such traits; heterozygous females do not. The characteristic features of X-linked recessive inheritance are:

1. Affected males are more common than affected females.
2. Affected fathers always transmit the allele to daughters, most of whom are unaffected carriers.
3. Affected fathers never transmit the trait to sons.
4. On average, half of the sons of a heterozygous mother are affected.
5. A woman will be affected only if her father was affected.

Examples of X-linked recessive inheritance include colorblindness, muscular dystrophy, and two forms of hemophilia.

Traits caused by dominant alleles of genes located on the X chromosome exhibit **X-linked dominant** inheritance. Hemizygous males and homozygous or heterozygous females exhibit such traits. The characteristic features of X-linked dominant inheritance are:

1. Affected males always have affected mothers.
2. All daughters of affected males are affected.
3. On average, half the sons and half the daughters of a heterozygous mother are affected.

Examples of X-linked dominant inheritance include vitamin-D-resistant rickets and focal dermal hypoplasia.

When attempting to decipher the most likely mode of inheritance from pedigree information, you should first hypothesize a certain mode of inheritance (for example, sex-linked recessive). Then, examine the pedigree and establish whether the information in the pedigree is consistent with the hypothesis. If the data are not consistent, then the hypothesis is wrong. Try all hypotheses until you identify the ones that are consistent with all the information. Several useful steps in analyzing a pedigree are:

1. Establish if possible whether the trait is common or rare in the general population. For rare traits, unrelated unaffected individuals are almost always homozygous normal. For recessive traits that are common in the general population, however, unrelated unaffected individuals might be heterozygous. This possibility must be considered for common traits.
2. Establish whether the trait is dominant or recessive. Are generations skipped? Dominant traits do not skip generations. Recessive traits usually skip generations, especially if they are rare in the general population. If unaffected parents have affected children, the trait must be recessive.
3. Examine whether both males and females are affected. If only males are affected, X-linked recessive inheritance is a good candidate. Males affected by X-linked recessive traits always have homozygous or heterozygous mothers and daughters.
4. Examine whether fathers pass the trait to sons. If they do, the trait cannot be X-linked.
5. Examine whether fathers *always* pass the allele to daughters. If they do, X-linked inheritance is a good candidate. If they do not, X-linked inheritance is ruled out.
6. Establish whether any matings are consanguineous (occurring between relatives). Such traits are usually recessive and occur more frequently among the children of related parents than among those of unrelated parents.

Self-Testing Questions

1. Six small pedigrees are shown below.

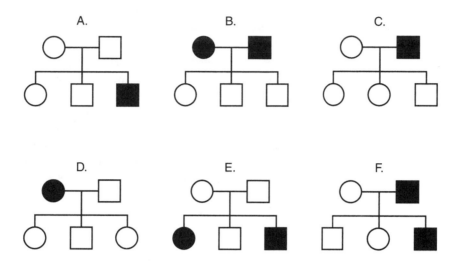

Indicate with a Y (Yes) or N (No) whether each of the pedigrees is consistent with each of the following modes of inheritance. Assume that these traits are common in the general population.

	A	B	C	D	E	F
Autosomal recessive	___	___	___	___	___	___
Autosomal dominant	___	___	___	___	___	___
X-linked recessive	___	___	___	___	___	___
X-linked dominant	___	___	___	___	___	___
Y-linked	___	___	___	___	___	___

➤ Solutions

Answer

	A	B	C	D	E	F
Autosomal recessive	Y	N	Y	Y	Y	Y
Autosomal dominant	N	Y	Y	Y	N	Y
X-linked recessive	Y	N	Y	N	N	Y
X-linked dominant	N	N	N	Y	N	N
Y-linked	N	N	N	N	N	N

Pedigree A is not consistent with any form of dominant or Y-linked inheritance, because the parents are unaffected. This leaves autosomal recessive and X-linked recessive. If both parents were heterozygous, pedigree A is consistent with autosomal recessive inheritance. If only the mother were heterozygous, pedigree A is consistent with X-linked recessive inheritance.

Pedigree B is not consistent with any form of recessive or Y-linked inheritance, because none of the children are affected. This leaves autosomal dominant and X-linked dominant. If both parents were heterozygous, pedigree B is consistent with autosomal dominant inheritance. In this case, none of the children received a dominant allele. Pedigree B is not consistent with X-linked dominant inheritance. The affected father contributed his X chromosome to his daughter, yet the daughter is unaffected.

Pedigree C is not consistent with X-linked dominant inheritance. The affected father contributed his X chromosome to his daughters, yet they are unaffected. Pedigree C is not consistent with Y-linked inheritance, because the affected father does not contribute the trait to his son. This leaves autosomal recessive, X-linked recessive, and autosomal dominant. Pedigree C is consistent with all three of these modes of inheritance. In the case of autosomal or X-linked recessive, the mother could either be heterozygous or homozygous normal. In the case of autosomal dominant inheritance, the father is heterozygous but did not contribute the dominant allele to any of his children.

Pedigree D is not consistent with X-linked recessive, because the son, who received his X chromosome from his mother, is not affected. Pedigree D is not consistent with Y-linked inheritance, because the only affected individual is female. This leaves autosomal recessive, autosomal dominant, and X-linked dominant. Pedigree D is consistent with all three of these modes of inheritance. In the case of autosomal recessive, the father could either be heterozygous or homozygous normal. In the case of autosomal or X-linked dominant, the mother is heterozygous but did not contribute the dominant allele to any of her children.

Pedigree E is not consistent with any form of dominant or Y-linked inheritance, because the parents are unaffected. This leaves autosomal recessive and X-linked recessive. If both parents were heterozygous, pedigree E would be consistent with autosomal recessive inheritance. Pedigree E is not consistent with X-linked recessive; the affected daughter received an X chromosome from her father, yet he contains only a single X chromosome and is unaffected.

Pedigree F is not consistent with X-linked dominant inheritance, because the daughter, who received an X chromosome from her affected father, is unaffected. Pedigree F is not consistent with Y-linked inheritance, because the affected father does not transmit the trait to both sons. This leaves autosomal recessive, autosomal dominant, and X-linked recessive. Pedigree F is consistent with all three of these modes of inheritance. In the case of either autosomal or X-linked recessive, the mother is heterozygous. In the case of autosomal dominant, the father is heterozygous and contributes the dominant allele to only one child.

2. Indicate whether the following statements are true or false.

_____ (a) Mothers transmit X chromosomes to both their sons and daughters.

_____ (b) Fathers transmit X chromosomes to their sons, but not to their daughters.

_____ (c) For X-linked dominant traits, affected daughters always have affected fathers.

_____ (d) For autosomal dominant traits, affected individuals have at least one affected parent.

_____ (e) If both parents are affected by an autosomal recessive trait, all children will be affected.

_____ (f) For X-linked dominant traits, the number of affected females in a population is approximately equal to the number of affected males.

_____ (g) Autosomal traits generally affect equal numbers of males and females.

_____ (h) For X-linked recessive traits, the number of carrier females in a population is approximately equal to the number of affected males.

_____ (i) For X-linked dominant traits, affected sons always have affected mothers.

_____ (j) If a child and both parents are affected by a genetic trait, the trait is dominant.

Answer

_T_ (a) Mothers transmit X chromosomes to both their sons and daughters.

_F_ (b) Fathers transmit X chromosomes to their sons, but not to their daughters.

_F_ (c) For X-linked dominant traits, affected daughters always have affected fathers.

_T_ (d) For autosomal dominant traits, affected individuals have at least one affected parent.

_T_ (e) If both parents are affected by an autosomal recessive trait, all children will be affected.

_F_ (f) For X-linked dominant traits, the number of affected females in a population is approximately equal to the number of affected males.

_T_ (g) Autosomal traits generally affect equal numbers of males and females.

_F_ (h) For X-linked recessive traits, the number of carrier females in a population is approximately equal to the number of affected males.

_____T_____   **(i)** For X-linked dominant traits, affected sons always have affected mothers.

_____F_____   **(j)** If a child and both parents are affected by a genetic trait, the trait is dominant.

3. Only one of the following four pedigrees is *not* consistent with X-linked dominant inheritance. Which is it? Affected individuals are shaded.

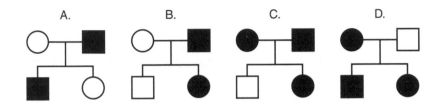

A.   **B.**   **C.**   **D.**

Answer A

4. The following pedigree concerns a trait that is common in the general population. Affected individuals are shaded.

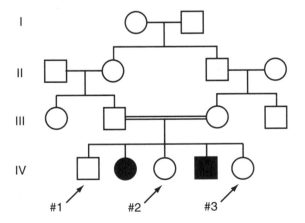

(a) What is the most likely mode of inheritance?

(b) How many individuals in this pedigree *must* be heterozygous? Which are they?

(c) What is the probability that male #1 (IV.1) is heterozygous?

(d) What is the probability that individuals #1, #2, and #3 (IV.1, IV.3, and IV.5) are all homozygous normal?

➤ *Solutions*

(a) *Answer:* Autosomal recessive

Any form of dominant inheritance is excluded, because affected children have unaffected parents. X-linked recessive inheritance is excluded because the affected daughter (IV.2) has an unaffected father. Daughters receive one of their X chromosomes from their father. If this were X-linked recessive inheritance, IV.2 is homozygous. Yet her father, from whom she received an X chromosome, is unaffected. Thus, X-linked recessive inheritance is excluded, leaving only autosomal recessive inheritance.

(b) *Answer:* Four. They are III.2, III.3, either II.1 or II.2, and either II.3 or II.4.

Both parents of the affected individuals must be heterozygous (III.2 and III.3). From whom did they inherit their autosomal recessive alleles? Clearly, they inherited them from their parents, but from which ones? It is not possible to be certain. The carrier male of generation III (III.2) might have inherited his allele from either his mother or father. Similarly, the carrier female of generation III (III.3) might have inherited her allele from either her mother or father. It is uncertain whether the matriarch or patriarch of this family (I.1 and I.2) are heterozygous. One or both of them *might* be heterozygous, but individuals II.1 and II.4 might have introduced the recessive alleles to the family. Thus, at least four individuals in this pedigree are heterozygous (III.2, III.3, either II.1 or II.2, and either II.3 or II.4).

(c) *Answer:* ⅔

In parts **(a)** and **(b)** of this question, it was established that this trait is inherited as an autosomal recessive and that both parents of male #1 are heterozygous carriers. Two heterozygous parents yield a 1:2:1 ratio of children that are affected homozygotes, heterozygotes, and normal homozygotes, respectively. Male #1 is not affected by the trait. Thus, he is either heterozygous or homozygous normal. Considering only *unaffected* children, ⅔ of them are expected to be heterozygous and ⅓ of them are expected to be homozygous normal.

(d) *Answer:* $\frac{1}{27}$

> In part **(c)** of this question, it was established that each unaffected child of generation IV has a $\frac{1}{3}$ probability of being homozygous normal. When calculating the combined probability that all three unaffected children are homozygous normal, each child represents an independent trial. Thus, the Product Rule applies. The probability that all three children are homozygous normal is $\frac{1}{3} \times \frac{1}{3} \times \frac{1}{3} = \frac{1}{27}$.

5. Focal dermal hypoplasia (FDH) is an X-linked dominant disease characterized by skin lesions and developmental defects in heterozygous females. FDH is lethal in hemizygous males, affected individuals dying as early embryos *in utero*. What proportion of the liveborn offspring of mothers affected by FDH exhibit the disease?

➤ **Solution**

Answer: $\frac{1}{3}$

Let X^{FDH} indicate an X chromosome carrying the dominant FDH allele and X^+ indicate an X chromosome carrying the normal allele. A heterozygous mother (genotype $X^{FDH}X^+$) produces 2 types of ova (X^{FDH} and X^+). Normal males produce 2 types of sperm (X^+ and Y). Random union of sperm and ova yield 4 genotypes of offspring ($X^{FDH}X^+$, X^+X^+, $X^{FDH}Y$, and X^+Y). Males of genotype $X^{FDH}Y$ die as early embryos; they are absent from liveborn children. Thus, $\frac{1}{3}$ of liveborn offspring are affected by FDH. All affected children are girls.

6. The following pedigree concerns a sex-linked recessive trait that is common in the general population.

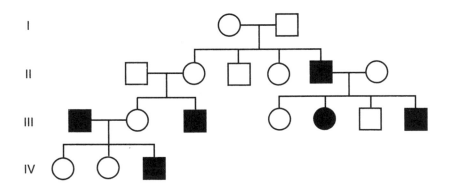

How many individuals in this pedigree must be heterozygous? Which are they?

Answer

Seven females must be heterozygous (I.1, II.2, II.6, III.2, III.4, IV.1, and IV.2).

7. In the following pedigree, affected individuals are shaded.

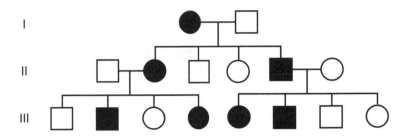

(a) Assuming that this trait is dominant, is it autosomal dominant or X-linked dominant?

(b) Assuming that this trait is a common recessive trait, is it autosomal recessive or X-linked recessive?

➤ **Solutions**

(a) *Answer:* Autosomal dominant

Much of this pedigree is consistent with X-linked dominant inheritance, but there is a key part that is not. Daughters always receive an X chromosome from their fathers. The affected male of generation II (II.5) has two daughters, one of whom is affected and one of whom is unaffected. Thus, this pedigree is not consistent with X-linked dominant inheritance. On the other hand, this pedigree is completely consistent with autosomal dominant inheritance, in which case all affected individuals are heterozygous.

(b) *Answer:* Autosomal recessive

This pedigree is consistent with autosomal recessive, but not with X-linked recessive, inheritance. X-linked recessive inheritance is excluded because affected mothers (I.1 and II.2) have both affected and unaffected sons. If this were X-linked recessive, affected mothers would be homozygous. Sons receive their X chromosome from their mother. Thus all sons of affected mothers are affected for X-linked recessive traits, which is not true in this pedigree. On the other hand, this pedigree is completely consistent with inheritance of a common autosomal recessive trait, in which case individuals I.2, II.1, and II.6 are heterozygous.

8. You are trying to decide if the following pedigree is caused by inheritance of a genetic trait or by non-genetic influences. Affected individuals are shaded.

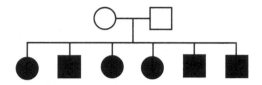

   (a) If this condition were caused by a single-gene genetic trait, what is the mode of inheritance?
   (b) Given your answer to part (a), what is the probability that 2 unaffected parents would have 6 affected offspring?

   Answer
   (a) Autosomal recessive
   (b) $(\frac{1}{4})^6 = \frac{1}{4096}$

9. The following pedigree concerns a trait that is common in the general population. Affected individuals are shaded.

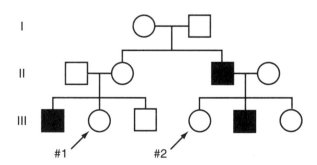

   (a) Assuming that this is a sex-linked recessive trait, what is the probability that female #1 (III.2) is heterozygous? Female #2 (III.4)?
   (b) Assuming that this is an autosomal recessive trait, what is the probability that female #1 (III.2) is heterozygous? Female #2 (III.4)?

> *Solutions*

(a) *Answer:* Female #1 = ½; female #2 = 1

To calculate the probability that female #1 is heterozygous, we must first calculate the probability that her mother (II.2) is heterozygous. If she is, then the probability that female #1 inherited an affected X chromosome is ½. What is the probability that II.2 is heterozygous? II.2 had an affected son. Thus, she *must* be heterozygous (probability = 1). It follows then that the probability of III.2 (female #1) being heterozygous is ½. What about female #2 (III.4)? She has an affected father. Fathers contribute their single X chromosome to all daughters. Thus, female #2 *must* be heterozygous (probability = 1).

(b) *Answer:* Female #1 = ⅔; female #2 = 1

To calculate the probability that female #1 is heterozygous, we must first calculate the probability that *both* her parents are heterozygous. Female #1 has an affected brother. Thus, both parents *must* be heterozygous. In a cross between 2 heterozygotes, the offspring are a 1:2:1 ratio of genotypes (homozygous recessive:heterozygous:homozygous dominant). Heterozygotes constitute ½ of total offspring, but they constitute ⅔ of *unaffected* offspring. Because female #1 is not affected, the probability that she is heterozygous is ⅔. What about female #2? She received 1 chromosome from her father, who is affected and therefore homozygous. Thus, female #2, who is not affected, *must* be heterozygous (probability = 1).

10. Only one of the following four pedigrees is consistent with autosomal recessive, sex-linked recessive, autosomal dominant, and X-linked dominant inheritance (all four). Which is it? Affected individuals are shaded.

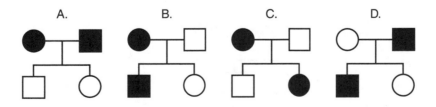

Answer B

11. The following pedigree concerns a trait that is common in the general population. Affected individuals are shaded.

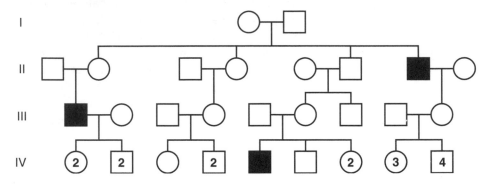

With which of the following modes of inheritance is this pedigree consistent? Check all that apply.

___ Autosomal recessive
___ Autosomal dominant
___ X-linked recessive
___ X-linked dominant

Answer

_X_ Autosomal recessive
___ Autosomal dominant
_X_ X-linked recessive
___ X-linked dominant

12. On the following blank pedigree diagram, shade one individual such that the resulting pattern of inheritance is consistent with a common sex-linked dominant trait. Shade exactly one square or circle to indicate an affected individual.

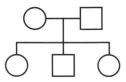

Answer

There is only one correct answer. The mother is affected.

13. The following pedigree concerns inheritance of an autosomal recessive trait that is common in the general population. Affected individuals are shaded.

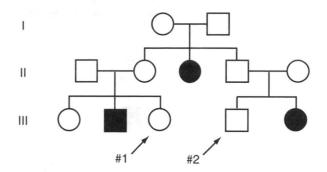

(a) What is the minimal number of individuals in this pedigree that are heterozygous?

(b) What is the probability that female #1 (III.3) is heterozygous?

(c) If female #1 (III.3) marries male #2 (III.4), what is the probability that their first child will be affected?

Answer
(a) 6
(b) ⅔
(c) ⅑

14. Is this a pedigree of autosomal dominant or X-linked dominant inheritance? Affected individuals are shaded. Individuals implied but not shown in the pedigree are unaffected.

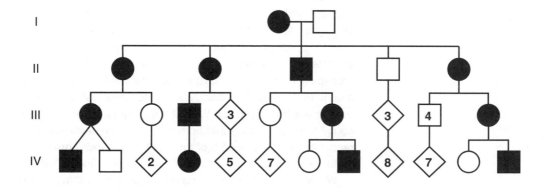

Answer Autosomal dominant

15. The pedigree below shows inheritance of Tay-Sachs disease, an autosomal recessive trait, in two unrelated families. The indicated couple (II.2 and II.3) would like to have children. What is the probability that, if the couple has 3 children, none of them will be affected by Tay-Sachs disease?

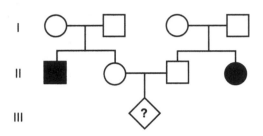

➤ **Solution**
   Answer: $(\frac{2}{3})^2(\frac{3}{4})^3 + \frac{5}{9} = \frac{107}{144}$

To calculate the probability that the couple will have affected (or unaffected) children, we must first calculate the probability that the parents are both heterozygous. The couple will have affected children only if they are both heterozygous. The prospective mother (II.2) has an affected brother. Thus, both of her parents are heterozygous. The prospective father (II.3) has an affected sister. Thus, both of his parents are heterozygous. In a cross between 2 heterozygotes, the offspring are a 1:2:1 ratio of genotypes (homozygous recessive:heterozygous:homozygous dominant). Heterozygotes constitute $\frac{1}{2}$ of total offspring, but they constitute $\frac{2}{3}$ of *unaffected* offspring. Because the prospective mother is not affected, the probability that she is heterozygous is $\frac{2}{3}$. Similarly, the probability that the prospective father is heterozygous is also $\frac{2}{3}$. The probability that they are *both* heterozygous is $(\frac{2}{3})(\frac{2}{3}) = \frac{4}{9}$. If both parents are heterozygous, what is the probability that they will have unaffected children? From a cross between 2 heterozygotes, $\frac{1}{4}$ of the children are expected to be affected and $\frac{3}{4}$ are expected to be unaffected. The probability that two heterozygous parents will have 3 unaffected children is $(\frac{3}{4})^3 = \frac{27}{64}$. This calculation concerns situations in which both parents are heterozygous. What if the parents are not both heterozygous? If the probability that both parents are heterozygous is $\frac{4}{9}$, the probability that one or both of them is homozygous normal is $1 - \frac{4}{9} = \frac{5}{9}$. In this case ($p = \frac{5}{9}$), *all* of their children are unaffected.

We can now calculate the combined probability that the couple will have 3 unaffected children. If both parents are heterozygous ($p = \frac{4}{9}$), the probability of having 3 unaffected children is $\frac{17}{64}$. If one or both parents is not heterozygous ($p = \frac{5}{9}$), the probability of having 3 unaffected children is 1. The combined probability of these mutually exclusive outcomes is $(^4\!/_9)(^{27}\!/_{64})+(^5\!/_9)(1) = {}^{107}\!/_{144}$.

---

16. The following pedigree describes inheritance of Wilson's disease, a neurodegenerative disease associated with defects in copper metabolism. Equal numbers of males and females are affected by Wilson's disease. What is the most likely mode of inheritance?

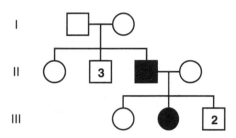

Answer Autosomal recessive

17. Charcot-Marie-Tooth neuropathy is inherited as an X-linked dominant disease. Pituitary dwarfism is inherited as an autosomal dominant disease. Of the following two pedigrees, which describes Charcot-Marie-Tooth neuropathy and which describes pituitary dwarfism? Affected individuals are shaded.

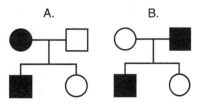

Answer

Inheritance of Charcot-Marie-Tooth neuropathy is described in pedigree A; inheritance of pituitary dwarfism is described in pedigree B.

18. The following pedigree concerns a trait that is rare in the general population. Affected individuals are shaded.

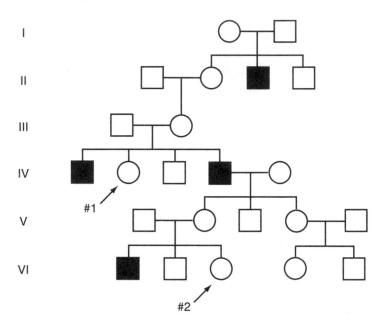

(a) What is the most likely mode of inheritance?

(b) How many individuals in this pedigree *must* be heterozygous? Which are they?

(c) What is the probability that female #1 (IV.2) is heterozygous?

(d) If female #2 (VI.3) marries a normal unrelated man, what are the chances that their first child will be affected?

➤ *Solutions*

(a) *Answer:* X-linked recessive

This pedigree is not consistent with any form of dominant inheritance, because affected children often do not have affected parents. This leaves autosomal and X-linked recessive inheritance. While both are formally possible, X-linked recessive is more likely for two reasons. First, only males are affected. This is a classic feature of X-linked recessive inheritance. Second, if this inheritance were autosomal recessive, then individuals I.1, I.2, III.1, III.2, IV.5, V.1, and V.2 must be heterozygous. Of these obligate heterozygotes, males III.1 and V.1 married into the family. That is, they would have to be heterozygous, yet they did not inherit the recessive allele from the pedigree itself. (They must have inherited it from their own parents.) This is unlikely for a trait that is rare in the general population. Such information does not completely *disprove* autosomal recessive inheritance, but it makes it unlikely. On the other hand, this pedigree is completely consistent with X-linked recessive inheritance. Each affected male has a heterozygous mother, and all heterozygous females inherited the recessive allele, ultimately, from the matriarch of the family (I.1).

(b) *Answer:* 5 females must be heterozygous. They are individuals I.1, II.2, III.2, V.2, and V.4.

For X-linked recessive traits, the mothers and daughters of affected males are always either heterozygous or homozygous. Thus, females I.1, III.2, V. 2, and V.4 are obligate heterozygotes. In addition, female II.2, who contributed the allele to III.2, is also heterozygous.

(c) *Answer:* ½

Female #1 (IV.2) has a heterozygous mother. Thus, the probability that she inherited an X chromosome carrying the recessive allele is ½.

(d) *Answer:* ⅛

To calculate the probability that female #2 (VI.3) will have an affected child, we must first calculate the probability that she is heterozygous. Female #2 has an affected brother. Thus, her mother is certainly heterozygous. It then follows that the probability of female #2 being heterozygous is ½. If she is heterozygous, then ½ of her sons will be affected. We are now in a position to calculate the combined probability of having an affected child. The probability that female #2 is heterozygous is ½. The probability that her first child is a boy is ½. The probability that a boy will be affected is ½. Thus, the combined probability of her firstborn child being affected is (½)(½)(½) = ⅛.

19. The pedigree shown on p. 82 describes inheritance of an autosomal recessive disease that is rare in the general population. Affected individuals are shaded. Individuals IV.1 and IV.2 are second cousins. What is the probability that a child born to this couple will be affected by the disease?

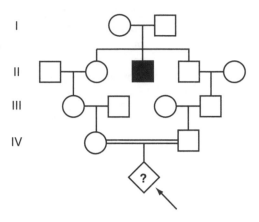

► **Solution**

Answer: $(\frac{2}{3})^2(\frac{1}{2})^4(\frac{1}{4}) = \frac{1}{144}$

To calculate the probability of an affected child, we must first calculate the probability that both of the parents (IV.1 and IV.2) are heterozygous. If they are heterozygous, then $\frac{1}{4}$ of their children are expected to be affected. What is the probability that IV.1 is heterozygous? Looking backwards through the pedigree, the only individuals that are *certain* to be heterozygous are the matriarch and patriarch of this family (I.1 and I.2). Because they have an affected son, they must be heterozygous. In a cross between 2 heterozygotes, the offspring are a 1:2:1 ratio of genotypes (homozygous recessive:heterozygous:homozygous dominant). Heterozygotes constitute $\frac{1}{2}$ of total offspring, but they constitute $\frac{2}{3}$ of *unaffected* offspring for an autosomal recessive trait. Thus, the probability that each of the 2 unaffected children of generation II (II.2 and II.4) are heterozygous is $\frac{2}{3}$. If they are heterozygous, the probability that an unaffected child inherits the recessive allele is $\frac{1}{2}$.

We are now in a position to calculate the combined probability that IV.1 and IV.2 are heterozygous. The probability that IV.1's grandmother (II.2) is heterozygous is $\frac{2}{3}$. If she is, the probability that IV.1's mother (III.1) is heterozygous is $\frac{1}{2}$. If she is, then the probability that IV.1 is heterozygous is $\frac{1}{2}$. All of these independent events must occur in order for IV.1 to be heterozygous. Thus, the combined probability that IV.1 is heterozygous is $(\frac{2}{3})(\frac{1}{2})(\frac{1}{2})$. Similarly, the combined probability that her husband (IV.2) is heterozygous is $(\frac{2}{3})(\frac{1}{2})(\frac{1}{2})$. The probability that they are *both* heterozygous is $(\frac{2}{3})^2(\frac{1}{2})^4 = \frac{1}{36}$.

If both parents are heterozygous, then $\frac{1}{4}$ of their children are expected to be affected. Thus, the probability that this couple will have an affected child is $(\frac{1}{36})(\frac{1}{4}) = \frac{1}{144}$.

20. The following pedigree concerns a sex-linked recessive trait that is common in the general population. Affected individuals are shaded.

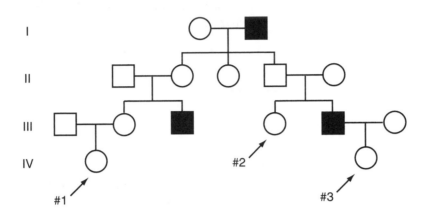

(a) How many individuals in this pedigree must be heterozygous? Which are they?

(b) What is the probability that female #1 (IV.1) is heterozygous?

(c) What is the probability that female #2 (III.4) is heterozygous?

(d) If female #3 (IV.2) has 3 sons, what is the probability that none of them will be affected?

Answer

(a) 4 (II.2, II.3, II.5, and IV.2)

(b) ¼

(c) ½

(d) ⅛

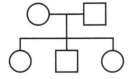

21. On the pedigree diagram at left, shade two individuals such that the resulting pattern of inheritance is consistent with a common sex-linked recessive trait. Shade exactly two individuals. Do not add individuals to the pedigree. Indicate only affected individuals (do not indicate heterozygotes).

Answer

There are four correct answers:

1. The mother plus the son are affected.

2., 3., and 4. The father plus any one child are affected.

22. On the following blank pedigree diagram, shade three individuals such that the resulting pattern of inheritance is consistent with a common sex-linked recessive trait. Shade exactly three individuals. Do not add individuals to the pedigree. Indicate only affected individuals (do not indicate heterozygotes).

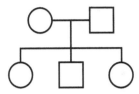

Answer

There are three correct answers: The father plus any two children are affected.

23. The pedigree shown below describes inheritance of adenosine deaminase (ADA) deficiency in two unrelated families. ADA deficiency is rare and is inherited as an autosomal recessive. The indicated couple (III.2 and III.3) would like to have children. What is the probability that, if they have two children, neither of them will be affected by ADA deficiency?

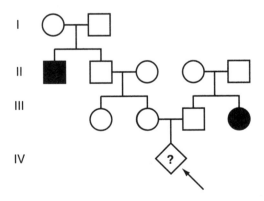

Answer $(2/3)^2(1/2)(3/4)^2 + (7/9) = 65/72$

24. Only one of the following four pedigrees is consistent with X-linked recessive inheritance. Which is it? Affected individuals are shaded.

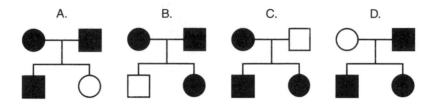

A. B. C. D.

Answer D

25. "Red-green colorblindness" is actually two distinct genetic traits, one affecting green color perception and another affecting red color perception. Both are X-linked recessive traits, but they result from alleles of two separate genes. Rare females are known who contain an allele for green colorblindness on one X chromosome, and an allele for red colorblindness on the other X chromosome. Because both the green and red colorblindness alleles are recessive, such females have normal color vision. If such a female marries a male with normal color vision, what proportion of their sons and daughters will be colorblind?

Answer

All daughters will have normal color vision. All sons will be colorblind; ½ will be green colorblind and ½ will be red colorblind.

26. The pedigree shown on p. 86 describes inheritance of Lesch-Nyhan syndrome, an X-linked recessive disease. Affected individuals are shaded. What is the probability that the indicated child (IV.1) will be affected by Lesch-Nyhan syndrome?

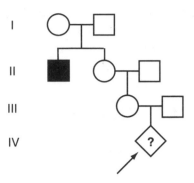

➤ **Solution**

Answer: $\frac{1}{16}$

To calculate the probability that IV.1 is affected, we must first calculate the probability that his or her mother is heterozygous. The great grandmother (I.1) is certainly heterozygous, because she had an affected son. The probability that the grandmother (II.2) is heterozygous is $\frac{1}{2}$. If she is heterozygous, the probability that the mother (III.1) is heterozygous is $\frac{1}{2}$. Thus, the combined probability that the mother (III.1) is heterozygous is $(\frac{1}{2})(\frac{1}{2}) = \frac{1}{4}$. If she is heterozygous, then $\frac{1}{2}$ of her sons will be affected by Lesch-Nyhan syndrome.

We are now in a position to calculate the combined probability of having an affected child. The probability that the mother is heterozygous is $\frac{1}{4}$. The probability that the indicated child (IV.1) is a boy is $\frac{1}{2}$. If a boy, the probability that he is affected is $\frac{1}{2}$. Thus, the combined probability of IV.1 being affected is $(\frac{1}{4})(\frac{1}{2})(\frac{1}{2}) = \frac{1}{16}$.

Linkage
and Recombination

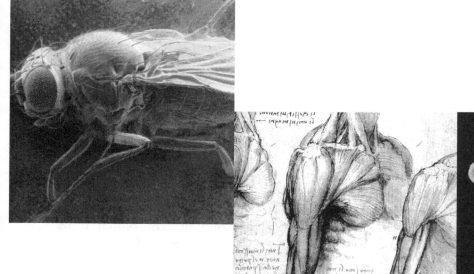

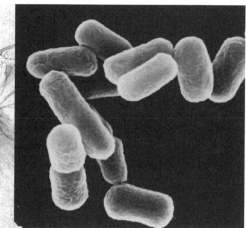

Summary

In general, the Principle of Independent Assortment applies only to genes located on different (nonhomologous) chromosomes. Two genes that are located on the same chromosome will tend to be transmitted together and are said to be linked. Recombinant chromosomes carrying new allelic combinations of **linked** genes are generated when a **crossover** occurs between the two genes. Crossing over involves the physical exchange of corresponding chromosome segments between homologues and takes place during the first meiotic prophase when homologous pairs of chromosomes engage in precise pairing or **synapsis** along their entire length. At this (**four-strand** or **tetrad**) stage, each chromosome has already replicated and consists of two sister chromatids. The meiotic products resulting from a single crossover event are two complementary recombinant chromatids and two complementary nonrecombinant (or parental) chromatids. The occurrence of crossing-over further enhances the ability of meiosis to generate genetic variability among gametes beyond that achievable from independent assortment alone.

Two genes that are located far apart on the same chromosome will be separated by crossing over more frequently than two genes that are very closely linked. Consequently, recombination frequencies can be used as a measure of the distance separating two linked genes. Distances are measured in terms of **map units** or **centimorgans** where 1% recombination is equal to 1 map unit or 1 centimorgan. Because genes are arranged in a linear sequence along chromosomes and because (over short intervals) map distances are additive, it is possible to use recombination frequencies to construct a **linkage map** that indicates the order and relative locations of a collection of linked genes. Linkage maps are extremely useful in predicting the outcome of various crosses and in the molecular identification of genes of interest.

Linkage mapping requires three separate crosses, each involving a different pair of genes, to determine the order and map distances for a set of three linked genes. The same information can be derived from a single cross involving all three genes. The triple heterozygotes used in such **three factor–crosses** generate eight distinct genotypic classes of gametes, including noncrossovers, single crossovers in the first marked interval, single crossovers in the second marked interval, and double crossovers. The noncrossover or parental types are always the most frequent class of gametes and the double crossovers are always the least frequent. From examination of the genetic composition of the parental and double crossover classes, it is possible to determine which of the three genes is in the middle. The observed frequency of double crossovers is usually less than the expected frequency if each crossover were an independent event. Instead, the occurrence of one crossover decreases the probability of another crossover taking place nearby. This phenomenon is referred to as **interference**.

As the distance between two linked genes becomes larger, the observed recombination frequency between them increasingly underestimates the true map distance, owing to the occurrence of undetected double crossover events within the interval. Therefore, the most accurate linkage maps are built up piece-meal from measurements of recombination frequencies over short intervals within which double recombinants are very rare. For genes that are very far apart on the same chromosome the maximum observable frequency of recombination is 50%, which is the equivalent of independent assortment.

In most higher organisms the gametes recovered in a cross are derived from many separate meioses. Consequently, we cannot directly determine the genetic composition of each of the four chromatids that were part of a single meiotic tetrad, and we cannot infer the sequence of meiotic events that produced any particular gamete that is recovered. For example, a gamete containing a noncrossover chromosome may have been produced by a meiosis in which there were no crossovers, a single crossover, or even a double crossover. Without knowing the composition of the other three products resulting from the same meiosis, we cannot distinguish these possibilities. However, in some organisms, particularly certain fungi, all four chromatids that participated in a single meiosis can be recovered together in an ordered array called a **linear tetrad**. Because the ordered array of products in a linear tetrad exactly reflects the chromosome movements that occurred in the preceding meiotic divisions, it is possible to determine the precise sequence of meiotic events that produced all the chromatids contained in a given tetrad. The analysis based on this special circumstance is called **tetrad analysis**. Tetrad analysis can provide much information about the details of meiosis and crossing over that could not be obtained otherwise. For this reason, studies carried out in fungi have been of key importance in elucidating the molecular details of recombination.

Tetrad analysis also permits the recombinational mapping of a gene relative to its centromere. A meiosis in which a crossover does not occur between the gene and the centromere will give rise to a tetrad exhibiting a characteristic spore pattern called a **first-division segregation** pattern. A meiosis in which a crossover *does* occur between the gene and the centromere will give rise to a tetrad with a different characteristic spore pattern, referred to as a **second-division segregation** pattern. Hence, the relative frequency of tetrads showing a second-division segregation pattern versus a first-division segregation pattern is a measure of the frequency of crossing over between the gene and the centromere.

Self-Testing Questions

1. Are the following statements true or false?

_____   **(a)** A single chiasma, visible during meiosis, leads to the formation of 4 recombinant and 0 nonrecombinant gametes.

_____   **(b)** Crossing over occurs during prophase I of meiosis.

_____   **(c)** In the absence of crossing over, recombinant gametes are not produced.

_____   **(d)** If both genes *A* and *B* are unlinked to gene *C*, then gene *A* is unlinked to *B*.

_____   **(e)** For a set of 3-point linkage data, if the value of I (Interference) equals −1.0, then twice as many double crossovers are observed than would be expected.

_____   **(f)** If crossing over did not occur during meiosis, all genes would exhibit first division segregation.

_____   **(g)** The maximum frequency of second division segregation is ½.

Answer

_F_   **(a)** A single chiasma, visible during meiosis, leads to the formation of 4 recombinant and 0 nonrecombinant gametes.

_T_   **(b)** Crossing over occurs during prophase I of meiosis.

_F_   **(c)** In the absence of crossing over, recombinant gametes are not produced.

_F_   **(d)** If both genes *A* and *B* are unlinked to gene *C*, then gene *A* is unlinked to *B*.

_T_   **(e)** For a set of 3-point linkage data, if the value of I (Interference) equals −1.0, then twice as many double crossovers are observed than would be expected.

_T_   **(f)** If crossing over did not occur during meiosis, all genes would exhibit first division segregation.

_F_   **(g)** The maximum frequency of second division segregation is ½.

2. In Drosophila, the dominant X-linked mutation Bar eye (*B*) is 5 centimorgans distant from a recessive mutation, *car*, that causes carnation-colored eyes. Wild-type flies have red eyes. What phenotypes and how many of them are expected among 1,000 male offspring of the cross $B +/ + car$ females $\times$ $+ car/Y$ males?

➤ **Solution**

Answer:

> 475 Bar red
> 475 non-Bar carnation
> 25 Bar carnation
> 25 non-Bar red

Male Drosophila have a single X chromosome, which they inherit from their mother. The mother is heterozygous for both *Bar* and *carnation*. Thus, the mother's nonrecombinant gametes are genotypes $B +$ and $+car$, and her recombinant gametes are $B\ car$ and $++$. What proportion of each does she produce? The B and car genes are 5 centimorgans distant. Thus, 5% of the mother's X chromosomes are recombinant and 95% are nonrecombinant. Recombinant chromosomes are equal numbers of $B\ car$ and $++$ (2.5% each). Nonrecombinant chromosomes are equal numbers of $B +$ and $+ car$ (47.5% each). Thus, among 1,000 male offspring, the expectations are 475 Bar red, 475 non-Bar carnation, 25 Bar carnation, and 25 non-Bar red.

---

 3. From the following linkage data, construct a linkage map of genes $a, b, c,$ and d.

Genes	Two-Factor Map Distance (centimorgans)
d and *c*	10 cM
c and *a*	13 cM
a and *d*	3 cM
b and *c*	18 cM
b and *d*	8 cM

Answer

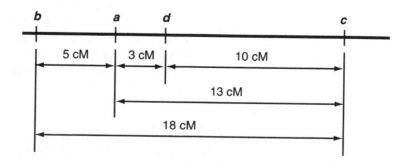

4. From the testcross (*abc*)/(+++) × (*abc*)/(*abc*), where the parentheses indicate that the order of genes *a*, *b*, and *c* is unknown, the following progeny were obtained.

Phenotype	Number
a ++	24
+*b* +	69
++*c*	2
ab +	4
a +*c*	75
+*bc*	20
abc	396
+++	410
Total:	1,000

Construct a linkage map for genes *a*, *b*, and *c*. Indicate all map distances.

➤ *Solution*
Answer:

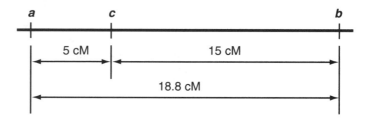

The first step in constructing the linkage map is calculating all of the two-factor map distances. Classes *a*++, +*b*+, *a*+*c*, and +*bc* result from crossovers between genes *a* and *b*; such offspring constitute 18.8% of total offspring. Classes *a*++, ++*c*, *ab*+, and +*bc* result from crossovers between genes *a* and *c*; such offspring constitute 5% of total offspring. Classes +*b*+, ++*c*, *ab*+, and *a*+*c* result from crossovers between genes *b* and *c*; such offspring constitute 15% of total offspring.

Knowing all two-factor map distances, the relative order of these genes is unambiguous. The order is *a-c-b* (or the reverse, *b-c-a*, which is equivalent). Genes *a* and *c* are 5 centimorgans distant, *c* and *b* are 15 centimorgans distant, and *a* and *b* are 18.8 centimorgans distant. Note that the calculated two-factor distance between genes *a* and *b* (18.8 cM) does not take into account the double crossovers (classes ++*c* and *ab*+). When such recombinants are considered, the distance between genes *a* and *b* becomes 20 cM.

5. The genes for both hemophilia A and red-green colorblindness are located on the X chromosome, with about 10% crossing

over between them. Linkage of a disease gene, such as hemophilia, to a relatively harmless one, like colorblindness, is
often used to predict the likely genotypes of offspring. The
pedigree shown below illustrates one such example.

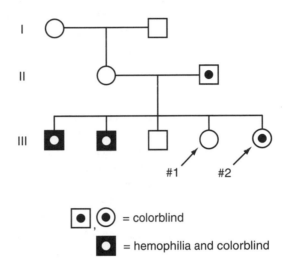

(a) If female #1 (III.4) marries a normal male, what proportion of her sons are expected to be affected with either
hemophilia or colorblindness? (Ignore the consequences
of crossing over in female #1 for this question, but do not
ignore the consequences of crossing over in female II.1.)

(b) If female #2 (III.5) marries a normal male, what proportion of her sons are expected to be affected with either
hemophilia or colorblindness?

➤ *Solutions*

(a) *Answer:* 50% colorblind (no hemophilia), 45% normal
(normal color vision, no hemophilia), 5% hemophilia
(normal color vision).

Let X^g, X^h, and X^+ indicate X chromosomes carrying a colorblindness allele, a hemophilia allele, and normal alleles of both
genes, respectively.

To predict the genotypes of female #1's offspring, we must first
deduce her genotype. Female #1 had a colorblind father. Thus,
one of female #1's X chromosomes carries a colorblindness allele.
Female #1 is not colorblind, so her second X chromosome must
either be X^+ or X^h. To decide whether female #1's second X

chromosome is X^+ or X^h, we must first deduce the genotype of her mother (II.1).

The mother (II.1) had two sons affected with both hemophilia and colorblindness. Because the RF between colorblindness and hemophilia is 10%, it is unlikely that both of these sons represent crossovers in the mother. A much more likely explanation is that one of the mother's X chromosomes contains *both* a colorblindness *and* a hemophilia allele (X^{gh}). The mother's second X chromosome must be normal (X^+), because she is unaffected by either trait. This is confirmed by the fact that she had both a normal son (III.3) and daughter (III.4). Thus, the mother's genotype is $X^{gh}X^+$.

Having established the mother's genotype, we can now deduce female #1's genotype. The only uncertainty concerns whether the X chromosome female #1 inherited from her mother is X^+ (nonrecombinant) or X^h (recombinant). (Female #1 is not colorblind, so it cannot be X^{gh} or X^g.) The mother's genotype is $X^{gh}X^+$. The RF between colorblindness and hemophilia is 10%. Thus, $p = 0.9$ that female #1 inherited an X^+ chromosome, and $p = 0.1$ that she inherited an X^h chromosome. We conclude that the genotype of female #1 is either $X^g X^+$ ($p = 0.9$) or $X^g X^h$ ($p = 0.1$).

If female #1 is genotype $X^g X^+$ ($p = 0.9$), half her sons are colorblind and none are affected by hemophilia. If female #1 is genotype $X^g X^h$ ($p = 0.1$), half her sons are colorblind and the other half are affected by hemophilia. Combining all this information, female #1's sons are expected to be:

50% colorblind (no hemophilia)
45% normal (normal color vision; no hemophilia)
5% hemophilia (normal color vision)

(b) *Answer:* All of her sons will be colorblind; 45% will be colorblind and have hemophilia.

In a manner parallel to that described above for part **(a)**, we deduce that the genotype of female #2 (III.5) is either $X^g X^{gh}$ ($p = 0.9$) or $X^g X^g$ ($p = 0.1$). If her genotype is $X^g X^{gh}$, all of her sons are colorblind and half of them have hemophilia. If her genotype is $X^g X^g$, all her sons are colorblind and none of them have hemophilia. Combining all this information, female #2's sons are expected to be:

55% colorblind (no hemophilia)
45% colorblind and hemophilia

6. The linkage map of a region of the Drosophila X chromosome is shown in the diagram on p. 97.

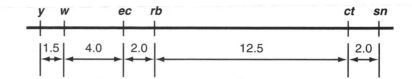

(a) Females of genotype *ec* +/+ *rb* are crossed with males of genotype *ec rb*/Y. What proportion of the male offspring are genotype *ec* +/Y?

(b) Females of genotype *ec* + *ct*/+ *rb* + are crossed with males of genotype *ec rb ct*/Y. Of 8,000 male offspring, 7 are genotype + + +/Y. What is the coefficient of coincidence for this region of the genetic map?

(c) Will the observed 2-factor map distance between *y* and *sn* be greater than or less than 22 centimorgans?

➤ *Solutions*

(a) *Answer:* 49%

Drosophila males contain a single X chromosome, which they inherit from their mothers. The mother is genotype *ec* +/+ *rb*. Thus, males of genotype *ec* +/Y result from a nonrecombinant X chromosome. What proportion of the mother's gametes contain this nonrecombinant X chromosome? The *ec* and *rb* genes are 2 centimorgans distant. Thus, 2% of the mother's gametes contain recombinant X chromosomes and 98% contain nonrecombinant X chromosomes. Half of the nonrecombinant X chromosomes are genotype *ec* +, and the other half are genotype + *rb*. Thus, 49% of the male offspring are expected to be genotype *ec* +/Y.

(b) *Answer:* C = 0.7

The coefficient of coincidence, C, equals the observed number of double crossovers divided by the expected number of double crossovers. Male offspring of genotype + + +/Y result from a double crossover in the mother. Seven such double crossovers are observed. To calculate C, we must first calculate the expected number of + + +/Y males, and then divide 7 by this number. How many + + +/Y males are expected?

The *ec* and *rb* genes are 2 cM distant; the *rb* and *ct* genes are 12.5 cM distant. Thus, the expected proportion of double crossovers (in the absence of interference) is (0.02)(0.125) = 0.0025. Among 8,000 gametes, (8,000)(0.0025) = 20 double crossovers are expected. Half of these are genotype + + +, and half are genotype *ec rb ct.* Thus, 10 double crossovers of geno-type + + + are expected. Seven are observed. Thus, the coeffi-cient of coincidence (C) equals $7/10$.

(c) *Answer:* less than 22 cM

The occurrence of undetected double crossovers causes the 2-factor map distance between *y* and *sn* to be less than 22 cM. What are undetected double crossovers? During meiosis of a *y sn*/+ + (or *y* +/+ *sn*) heterozygote, if *two* crossovers occur between the *y* and *sn* genes, the resulting gametes *appear* to be nonrecombi-nant even though two crossovers have occurred. The result is that the true frequency of recombination is underestimated. For genes that are far apart, undetected double crossovers occur in a signifi-cant proportion of meioses, and the magnitude of the underestima-tion is substantial. For genes that are close together, undetected double crossovers are rare and do not significantly affect the observed distances.

If all of the 2-factor map distances between *y* and *sn* are added together, the result is 22 cM. These individual values, however, were measured over short genetic distances, a situation in which undetected double crossovers are not a complication. Such values cannot be added together to estimate the 2-factor distance between *y* and *sn.* Undetected double crossovers cause such a calculation to be an overestimate of the observed 2-factor distance.

7. A plant of genotype *AA BB CC* is crossed with one of geno-type *aa bb cc,* where *A, B,* and *C* are dominant to *a, b,* and *c,* respectively. The F1 trihybrid is crossed with *aa bb cc.* The phenotypes and number of offspring obtained from this cross are shown in the table below.

Phenotype	Number of Offspring
ABC	200
aBC	200
Abc	200
abc	200
ABc	50
aBc	50
AbC	50
abC	50

Which, if any, of these genes (A,a, B,b, and C,c) are linked?

Answer

Genes B,b and C,c are linked to each other; gene A,a is unlinked to either B,b or C,c.

8. Crossing over occurs in female Drosophila, but it does not occur in males. Genes A,a and B,b are separated by 10 centimorgans on a Drosophilia autosomal chromosome. What genotypes are expected from the cross A B/a b females × A B/a b males, and in what proportions are they expected?

$$A \ B/a \ b \text{ females} \times A \ B/a \ b \text{ males}$$

Answer

$AB/ab = 0.45$
$AB/AB = 0.225$
$ab/ab = 0.225$
$AB/Ab = 0.025$
$AB/aB = 0.025$
$Ab/ab = 0.025$
$aB/ab = 0.025$

9. You examine ascospores of a *Neurospora* strain that is heterozygous for two genes, c and d. You observe the following asci:

									Number
Ascus type 1	cd	cd	cd	cd	++	++	++	++	54
Ascus type 2	cd	cd	++	++	++	++	cd	cd	46
Ascus type 3	cd	cd	++	++	cd	cd	++	++	56
Ascus type 4	++	++	++	++	cd	cd	cd	cd	48
Ascus type 5	++	++	cd	cd	cd	cd	++	++	49
Ascus type 6	++	++	cd	cd	++	++	cd	cd	47
								Total:	300

(a) Is gene c linked to gene d?
(b) Is either gene c or d linked to its centromere?

➤ **Solutions**

(a) *Answer:* Yes. No crossovers are detected among 300 asci.

Gene c is tightly linked to gene d. Among 2,400 (300 × 8) spores, not a single one results from crossing over between the c and d genes. Every spore that is genotype c is also genotype d, and every spore that is c^+ is also d^+. Although the genotype of the parent diploid is not stated, it is apparent that it is *cd*/++, and that crossing over between *c* and *d* was not observed.

(b) *Answer:* No. Both genes are unlinked to their centromere.

For genes that are unlinked to their centromere, two thirds of the asci exhibit second division segregation. Ascus types 2, 3, 5, and 6 (198 total) indicate second division segregation of both genes *c* and *d*. Such asci constitute approximately ⅔ of total asci. Thus, genes *c* and *d* are not linked to their centromere. Sufficient crossing over occurs between these genes and their centromere that they are unlinked.

10. You examine ascospores from a *Neurospora* strain that is heterozygous for two linked genes, *a* and *b*. You observe the following asci:

								Number
Ascus type 1	*ab*	*ab*	*ab*	*ab*	++	++	++	++	35
Ascus type 2	++	++	++	++	*ab*	*ab*	*ab*	*ab*	35
Ascus type 3	*ab*	*ab*	++	++	*ab*	*ab*	++	++	5
Ascus type 4	*ab*	*ab*	++	++	++	++	*ab*	*ab*	5
Ascus type 5	++	++	*ab*	*ab*	*ab*	*ab*	++	++	5
Ascus type 6	++	++	*ab*	*ab*	++	++	*ab*	*ab*	5
Ascus type 7	*ab*	*ab*	+*b*	+*b*	*a*+	*a*+	++	++	3
Ascus type 8	*ab*	*ab*	+*b*	+*b*	++	++	*a*+	*a*+	2
Ascus type 9	+*b*	+*b*	*ab*	*ab*	*a*+	*a*+	++	++	3
Ascus type 10	+*b*	+*b*	*ab*	*ab*	++	++	*a*+	*a*+	2
							Total:	100

Draw the linkage map of genes *a*, *b*, and their centromere. Indicate all map distances in centimorgans.

Answer

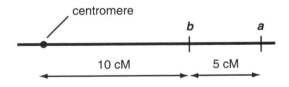

11. Dominant and recessive alleles of a gene in garden peas, *I* and *i*, cause seeds to be yellow and green, respectively. Dominant and recessive alleles, *F* and *f*, of another gene cause seeds to be spotted or not with violet color, respectively. A plant of genotype *II FF* is crossed with a plant of genotype *ii ff*, and the F1 are allowed to self-fertilize. The phenotypes and numbers of F2 progeny are shown in the table below.

Phenotype of F2	Number
yellow, spotted	89
green, spotted	31
yellow, not spotted	29
green, not spotted	11

Are the *I,i* and *F,f* genes linked?

Answer No

12. A female Drosophila has the following genotype on her X chromosome:

$$\frac{+\ +\ +\ +\ +\ +}{a\ b\ c\ d\ e\ f}$$

A recessive lethal mutation is located somewhere on the upper chromosome. (A recessive lethal mutation is one that kills both homozygous females and hemizygous males early in development; such animals are not recovered among the progeny of a cross. The lethal mutation is recessive to wild type, and the heterozygote above has normal viability.)

The heterozygous female was mated with wild-type males. The genotypes of 1,000 male offspring were:

Genotype	Number
abcdef	875
abcde +	30
abcd ++	10
abc +++	15
ab ++++	0
a +++++	0
++++++	0
+*bcdef*	15
++*cdef*	20
+++*def*	35
++++*ef*	0
+++++*f*	0
Total:	1,000

For simplicity, all multiple crossovers have been excluded from the data. The genes are shown in their proper order on the chromosome (*a* is on the left, *f* is on the right). Draw a genetic map for this chromosome. Locate the recessive lethal mutation on the map and indicate all of the 2-point distances (in centimorgans) between adjacent genes.

Answer

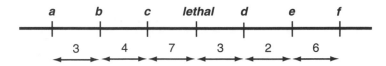

all distances in centimorgans

 13. The following genetic map describes three hypothetical human autosomal genes, each of which exhibits two alleles. Two-factor map distances are shown.

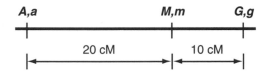

A = Artistic (dominant)
a = Inartistic (recessive)
M = Moral (dominant)
m = Immoral (recessive)
G = Generous (dominant)
g = Greedy (recessive)

Assume that these traits exhibit simple Mendelian dominance/recessiveness. The coefficient of coincidence for this map interval is 0.4.

An artistic, moral, generous heterozygous female of genotype *AMG/amg* marries an inartistic, immoral, greedy homozygous male of genotype *amg/amg*.
(a) What is the probability that their firstborn child will be inartistic, immoral, and greedy?
(b) What is the probability that their firstborn child will be inartistic, moral, and generous?
(c) What is the probability that their firstborn child will be artistic, immoral, and generous?

➤ **Solutions**
(a) *Answer:* $p = 0.354$

An inartistic, immoral, and greedy child inherits a gamete of genotype *amg* from his or her mother. This represents one of the mother's two nonrecombinant chromosomes. What proportion of the mother's chromosomes are nonrecombinant? Twenty percent of her gametes contain crossovers between genes *A* and *M,* and 10% contain crossovers between genes *M* and *G.* Some of her gametes, however, contain double crossovers, and such double crossovers contribute to the observed distances between *A, M,* and *G.* Thus, to precisely calculate how many nonrecombinant gametes the mother produces, we must first calculate the proportion of all single and double crossovers.

First, let's calculate the number of double crossovers. From the 2-factor map distances, the *expected* proportion of double crossovers is (0.2)(0.1) = 0.02. The coefficient of coincidence is 0.4. Thus, the *observed* proportion of double crossovers is (0.4)(0.02) = 0.008. Eight-tenths of one percent of the mother's gametes contain double crossovers.

Now we can calculate the proportion of single crossovers. The 2-factor distance between genes *A* and *M* is 20 cM. Thus, 20% of the mother's gametes contain crossovers between genes *A* and *M.* This figure (20%) includes all gametes that contain single crossovers plus all gametes that contain double crossovers (one between *A* and *M;* one between *M* and *G*). We've already calculated the proportion of gametes that contain double crossovers (0.8%). Thus, the proportion of the mother's gametes that contain single crossovers between *A* and *M* is 20% − 0.8% = 19.2%. Similarly, the proportion of the mother's gametes that contain single crossovers between *M* and *G* is 10% − 0.8% = 9.2%.

Putting all this information together, 19.2% + 9.2% + 0.8% of the mother's gametes contain either single or double crossovers. This leaves 70.8% of her gametes as nonrecombinant. Half of them are genotype *amg.* Thus, 35.4% of the children are expected to be inartistic, immoral, and greedy.

(b) *Answer:* $p = 0.096$

An inartistic, moral, generous child inherits a gamete of genotype *aMG* from his or her mother. This gamete results from a single crossover between genes *A* and *M* in the mother. In part (**a**), we calculated the proportion of the mother's gametes that contain single crossovers in this interval (19.2%). Half of these gametes are genotype *Amg,* and half are genotype *aMG.* Thus, 9.6% of the children are expected to be inartistic, moral, and generous.

(c) *Answer:* $p = 0.004$

An artistic, immoral, and generous child inherits a gamete of genotype *AmG* from his or her mother. This gamete results from a double crossover in the mother. In part **(a)**, we calculated the proportion of the mother's gametes that contain double crossovers (0.8%). Half of these gametes are genotype *AmG*, and half are genotype *aMg*. Thus, 0.4% of the children are expected to be artistic, immoral, and generous.

14. Recessive alleles *b, pr,* and *vg* cause black bodies, purple eyes, and vestigial wings, respectively, in Drosophila.

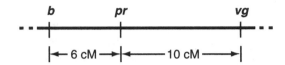

Females of genotype *b pr vg/b⁺ pr⁺ vg⁺* are crossed with males of genotype *b pr vg/b pr vg*. The phenotypes and numbers of offspring are shown in the table below.

Phenotype	Number
b⁺ pr⁺ vg⁺	4,224
b⁺ pr⁺ vg	500
b⁺ pr vg⁺	14
b⁺ pr vg	296
b pr⁺ vg⁺	280
b pr⁺ vg	10
b pr vg⁺	476
b pr vg	4,200
Total:	10,000

What is the coefficient of coincidence in this region?

Answer 0.4

15. Yeast genes *A,a* and *B,b* are tightly linked to each other. Both *A,a* and *B,b* are about 15 cM distant from gene *C,c*, but the order of *A,a* and *B,b* relative to *C,c* is unknown.

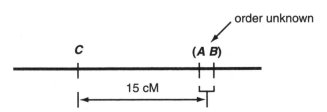

You sporulate a diploid of genotype $C(ab)/c(AB)$, where the parentheses indicate that the AB gene order is unknown. You analyze 1,000 random spores. The genotypes of those spores are shown below.

Genotype	Number
$C(ab)$	400
$c(AB)$	420
$C(AB)$	85
$c(ab)$	75
$C(Ab)$	8
$c(aB)$	11
$C(aB)$	1
$c(Ab)$	0
Total:	1,000

(a) What is the map distance in centimorgans between genes A,a and B,b?

(b) What is the correct order of the three genes?

➤ *Solutions*

(a) *Answer:* 2 cM

From a diploid of genotype $C(ab)/c(AB)$, haploid classes $C(Ab)$, $c(aB)$, and $C(aB)$, and $c(Ab)$ result from crossing over between genes A,a and B,b. Such classes constitute 2% of total haploids. Thus, genes A,a and B,b are 2 centimorgans apart.

(b) *Answer:* C-B-A

In a 3-factor cross, the least frequent class of offspring result from double crossovers. If the gene order were C-A-B then classes CAb and caB would result from double crossovers. If the gene order were C-B-A, then classes CBa and cbA would result from double crossovers. Classes $C(aB)$ and $c(Ab)$ are the least frequent. Thus, the gene order is C-B-A.

16. From a cross $(Xyz)/(xYZ) \times (xyz)/(xyz)$, where X,x, Y,y, and Z,z represent alleles of three linked genes whose relative order is unknown, the least frequent classes of progeny are genotype $(xyz)/(xyz)$ and $(XYZ)/(xyz)$. What is the correct relative order of genes X, Y, and Z?

Answer Y-X-Z (or the reverse, Z-X-Y)

17. A genetic map of a portion of chromosome 7 of maize (corn) is shown on p. 106.

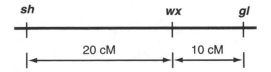

The coefficient of coincidence for this region of the genetic map is 0.5. From the cross *sh + gl/+ wx + × sh wx gl/sh wx gl*, how many offspring of genotype *sh wx gl/sh wx gl* are expected among 10,000 progeny?

Answer 50

18. You cross a haploid yeast of genotype *AB* with another of genotype *ab*, where *A,a* and *B,b* represent alleles of two different genes. You sporulate the diploid and dissect 200 asci. The following classes of asci are obtained.

	Genotypes of Spores	Number of Asci
Class 1	AB, AB, ab, ab	132
Class 2	AB, Ab, aB, ab	60
Class 3	Ab, Ab, aB, aB	8
	Total:	200

What is the map distance, in centimorgans, between the *A,a* and *B,b* genes? Do not correct for undetected double crossovers in your calculation.

➤ **Solution**
Answer: 19 cM

Ascus classes 1, 2, and 3 represent parental ditype (PD), tetratype (TT), and nonparental ditype (NPD) asci, respectively. Using the formula

$$\text{Map Distance (cM)} = \frac{(\frac{1}{2})(\text{TT}) + (\text{NPD})}{(\text{PD}) + (\text{TT}) + (\text{NPD})} \times 100$$

the map distance between genes *A,a* and *B,b* is calculated to be 19 centimorgans. Note that the formula used here does not correct for undetected double crossovers. After correcting for undetected double crossovers, the calculated map distance would be greater than 19cM.

19. In tomatoes, tall vines (*D*) is dominant to dwarf vines (*d*), and round fruit (*O*) is dominant to oval (*o*). The *D,d* and *O,o* genes are linked, with 15% crossing over between them.

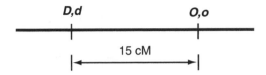

(a) You are given a tomato plant of unknown genotype that is tall and has round fruit. Your goal is to determine its genotype. You testcross this tomato plant with another whose genotype is *do/do* (dwarf; oval fruit). The F1 consist of:

88 tall plants having oval fruit
82 dwarf plants having round fruit
17 tall plants having round fruit
13 dwarf plants having oval fruit

What is the genotype of the tall plant with round fruit?

(b) If the tall plant with round fruit from Part **(a)** is allowed to self-fertilize, what proportion of its F1 will be dwarf plants with oval fruit?

Answer

(a) *Do/dO*

(b) 0.56%

20. Alleles *a*, *b*, and *c* are three recessive mutations of Drosophila. The *a*, *b*, and *c* genes are linked, but you don't know their order. You cross a heterozygous female of genotype (*a b +*)/(*+ + c*) with a male of genotype (*a b c*)/(*a b c*). (The parentheses indicate that the gene order is unknown.) The phenotypes of the F1 are:

Phenotype	Number
abc	55
ab+	333
a++	105
a+c	10
+bc	95
+b+	14
+++	65
++c	323
Total:	1,000

(a) What is the relative order of genes *a*, *b*, and *c*?

(b) Construct a linkage map for genes *a*, *b*, and *c*. Indicate the map distance between all adjacent genes.

Answer

(a) *b-a-c* (or the reverse, *c-a-b*)

(b)

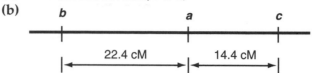

21. Both hemophilia (*h*) and favism (*gd*) are inherited as X-linked recessive traits. Hemophilia is an inherited disorder of blood clotting, and favism is an inherited hemolytic anemia caused by absence of the enzyme glucose-6-phosphate dehydrogenase. A phenotypically normal woman is known to have the X chromosome genotype *h* +/+ *gd*. The frequency of recombination between *h* and *gd* is 16%. What proportion of sons born to this woman are expected to be phenotypically normal with respect to both hemophilia and favism?

Answer 8%

22. The *Neurospora* genes *A,a* and *B,b* are linked to each other and to their centromere. You sporulate a diploid of genotype:

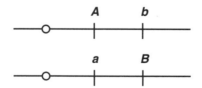

(a) For the following four ordered tetrads, indicate for each gene whether the pattern of spores indicates first or second division segregation. Enter "FDS" or "SDS."

									Gene *A,a*	Gene *B,b*
Ascus #1	*aB*	*aB*	*aB*	*aB*	*Ab*	*Ab*	*Ab*	*Ab*	_____	_____
Ascus #2	*Ab*	*Ab*	*ab*	*ab*	*AB*	*AB*	*aB*	*aB*	_____	_____
Ascus #3	*aB*	*aB*	*Ab*	*Ab*	*aB*	*aB*	*Ab*	*Ab*	_____	_____
Ascus #4	*aB*	*aB*	*ab*	*ab*	*Ab*	*Ab*	*AB*	*AB*	_____	_____

(b) Which gene (A,a or B,b) will exhibit the lowest frequency of second division segregation?

(c) The following diagrams show the above diploid at prophase I of meiosis . On each diagram, draw an "X" to indicate crossover(s) that yield the types of asci described below. Draw your Xs to indicate which chromatids are crossing over and where the crossover occurs relative to the centromeres and genes.

    i. Draw a crossover that yields second division segregation for both genes A,a and B,b.

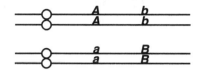

    ii. Draw a crossover that yields first division segregation for gene A,a and second division segregation for gene B,b.

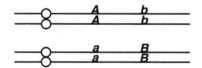

    iii. Draw a crossover that yields first division segregation for both genes A,a and B,b.

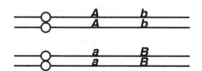

Answer

(a)

	Gene A,a	**Gene B,b**
Ascus #1	FDS	FDS
Ascus #2	SDS	FDS
Ascus #3	SDS	SDS
Ascus #4	FDS	SDS

(b) A,a

(c) *i.* Two chromatids (one from each homologue) cross over anywhere between the centromere and gene *A,a*.

ii. Two chromatids (one from each homologue) cross over anywhere between genes *A,a* and *B,b*.

iii. Two chromatids (one from each homologue) cross over anywhere to the left of the centromere or to the right of gene *B,b*; OR two sister chromatids cross over anywhere.

23. The following is a linkage map of *Neurospora*, which produces ordered ascospores following meiosis:

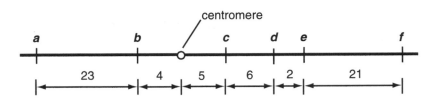

all distances in centimorgans

(a) Which gene exhibits the highest frequency of first division segregation?

(b) Which gene exhibits the lowest frequency of first division segregation?

(c) Which gene exhibits the highest frequency of second division segregation?

(d) If a diploid of genotype *c/+* is sporulated, what fraction of the asci exhibit second division segregation of the *c* gene?

(e) Which two genes are the most tightly linked?

(f) If a diploid of genotype *cd/++* is sporulated, what fraction of the asci exhibit first division segregation for gene *c* and second division segregation for gene *d*?

(g) What is the approximate frequency of recombination between genes *a* and *f*?

➤ **Solutions**

(a) *Answer:* Gene *b*

First division segregation occurs when crossing over does *not* occur between a gene and its centromere. Thus, the gene most tightly linked to its centromere (gene *b*) exhibits the highest frequency of first division segregation.

(b) *Answer:* Gene *f*

First division segregation occurs when crossing over does *not* occur between a gene and its centromere. Thus, the gene farthest away from its centromere (gene *f*) exhibits the lowest frequency of first division segregation.

(c) *Answer:* Gene *f*

Second division segregation results when crossing over occurs between a gene and its centromere. Thus, the gene farthest from its centromere (gene *f*) exhibits the highest frequency of second division segregation.

(d) *Answer:* 10%

Second division segregation (SDS) results when crossing over occurs between a gene and its centromere. The *c* gene is 5 cM from its centromere. Using the formula

$$\text{Map Distance (cM)} = \frac{(\frac{1}{2})(\text{SDS Asci})}{(\text{Total Asci})} \times 100,$$

the proportion of SDS asci is calculated to be 10%.

(e) *Answer:* Genes *d* and *e*

Genes *d* and *e*, separated by 2 centimorgans, are the most tightly linked.

(f) *Answer:* 12%

When crossing over occurs during meiosis, half of the spores in a single ascus are recombinant and half are nonrecombinant. Crossing over occurs between genes *c* and *d* in 6% of the spores. Thus, 12% of the asci contain spores in which genes *c* and *d* have crossed over.

(g) *Answer:* 50%

If all of the individual 2-factor distances between genes *a* and *f* are summed, the result is 61 cM. Two unlinked genes, however, exhibit a maximum frequency of recombination of 50%. Thus, the observed recombination frequency between genes *a* and *f* is 50%. They are sufficiently far apart on the same chromosome that they are unlinked.

Chromosome Aberrations

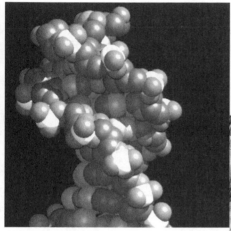

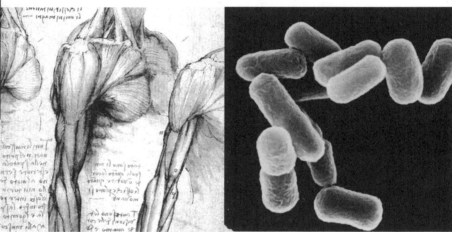

Summary

Chromosome aberrations can occur either spontaneously or following exposure to various agents such as ionizing radiation or certain chemicals. These aberrations are of two general types, those that alter chromosome number and those that alter chromosome structure. The various types of chromosome aberrations have significant effects on phenotype and on the transmission pattern of genes and are therefore of medical, agricultural, and evolutionary importance. Understanding the perturbations caused by these aberrations also reveals important insights into the normal properties of chromosomes.

Aneuploidy, the condition in which only one or a few chromosomes of a haploid set are missing or present in extra copies, arises from spontaneous or induced errors in mitosis or meiosis. The most common types of aneuploids are **monosomics**, in which one copy of a chromosome is lacking, and **trisomics**, in which there are three, rather than two, copies of a particular chromosome. The gene imbalance associated with aneuploidy has severe detrimental effects, including lethality, especially in animals. In humans, trisomy-21 causes Down syndrome. All other autosomal monosomies or trisomies in humans die before birth or shortly thereafter. The deleterious effects associated with aneuploidy demonstrate the exquisite genetic balance that normally occurs in most organisms.

Polyploidy also originates from errors in mitosis or meiosis and leads to the production of organisms containing three or more complete haploid sets of chromosomes. In nature, the occurrence of polyploidy is limited almost exclusively to the plant kingdom where it is very widespread. **Autopolyploids** contain extra chromosome sets all derived from the same species, whereas **allopolyploids** contain extra chromosome sets derived from two or more distinct but related species. In autopolyploids, the multiple copies of each chromosome often pair with each other as **multivalents** during meiosis. Because segregation from such multivalents is irregular, genetically unbalanced gametes that are nonfunctional are produced. For this reason, autopolyploids are generally sterile. In contrast, the chromosomes in allopolyploids generally pair as bivalents, leading to regular meiotic segregation. Consequently, allopolyploids are usually fertile. Both autopolyploids and allopolyploids are often larger and more vigorous than their progenitor diploid species. They may also display other novel qualities that make them commercially important.

Structural aberrations result from the breakage and reunion of chromosomes. An **inversion** results when a broken chromosome segment rotates through 180° before reattaching to the same chromosome with no loss or gain of chromosome material. If the centromere is included in the inverted segment, it is a **pericentric** inversion. If the centromere lies outside the inverted segment, it is a **paracentric** inversion. In inversion heterozygotes, when an inverted chromosome synapses with a standard chromosome during meiosis, the inverted segment forms a loop so that corresponding chromosome regions can pair. If a single crossover occurs in this region, the resulting chromatids are aberrant and are not recovered among the viable offspring. Hence, heterozygosity for an inversion prevents the recovery of recombinant chromatids within the inverted segment.

Translocations involve an interchange of segments between two nonhomologous chromosomes—again, with no loss or gain of chromosomal material. Meiotic pairing of homologous segments in a translocation heterozygote involves a complex of four chromosomes. Segregation from this complex can occur in several different ways, only one of which (**alternate segregation**) leads to the production of genetically balanced gametes and viable offspring. The other, equally frequent, segregation pattern (**adjacent segregation**) leads to unbalanced gametes and inviable offspring. Thus, an important consequence of translocation heterozygosity is semisterility. The problems associated with inversions and translocations result from their effects on the meiotic behavior of chromosomes. These chromosome rearrangements have little, if any, effects on the somatic phenotype of the organisms in which they occur. Instead, they affect the gametes produced in these organisms and the offspring resulting from these gametes.

In contrast with inversions and translocations, **deletions** and **duplications** involve, respectively, the loss or gain of a chromosome segment. Homozygosity for a deletion that includes more than one or two genes almost always causes lethality. Even in the heterozygous condition, large deletions are often associated with deleterious effects on the viability of the organism. The phenotypic effects of deletions are caused by gene imbalance associated with the loss of genetic information, but the meiotic behavior of deleted chromosomes is normal. Because deletions entirely lack a particular chromosome segment, any recessive alleles on the standard chromosome that fall within the region uncovered by a deletion will be expressed phenotypically in organisms heterozygous for the deletion. This result, together with the fact that the precise cytological limits of a deletion can often be determined by microscopic observation, enables use of deletions to map the physical location of genes on chromosomes.

Unless the extent of the extra segment present is very large, duplications will often have only modest phenotypic effects in either the homozygous or the heterozygous condition. However, duplications have long-term evolutionary importance by providing a repository of excess or redundant genes that are free to sustain mutations without harmful impact. The duplicated genes are therefore capable of evolving new genetic functions without loss of essential existing functions.

Self-Testing Questions

1. Indicate whether the following statements are true or false.
   (a) A pericentric inversion is one in which the centromere is not included in the inverted region.
   (b) A triploid cell is a euploid cell.
   (c) Fertile amphidiploids always have an even number of chromosomes.
   (d) Crossing over within the inverted region of a paracentric inversion heterozygote generates two dicentric chromosomes.
   (e) An amphidiploid derived from a cross between two diploid plants is a tetraploid plant.
   (f) Allotetraploids are tetraploids in which the chromosome sets derive from the same species.
   (g) A child with Down syndrome (trisomy-21) has a euploid karyotype.
   (h) Because of their unbalanced gametes, translocation heterozygotes are usually less fertile than either translocation homozygotes or strains not containing chromosome rearrangements.
   (i) An autotetraploid will have twice the number of linkage groups as a diploid of the same species.
   (j) When a monosomic cell undergoes meiosis, all four of the resulting gametes are aneuploid.

> *Solutions*

(a) *Answer:* _F_ A pericentric inversion is one in which the centromere is not included in the inverted region.

In *peri*centric inversions, the centromere is included in the inverted region. In *para*centric inversions, the centromere *is not* included in the inverted region.

(b) *Answer:* _T_ A triploid cell is a euploid cell.

Euploid cells contain any multiple of the monoploid number of chromosomes. Triploids contain three full haploid sets of chromosomes. Thus, a triploid cell is a euploid cell.

(c) *Answer:* _T_ Fertile amphidiploids always have an even number of chromosomes.

Amphidiploids are formed by doubling the number of chromosomes contained in an interspecific hybrid. Thus, amphidiploids always contain an even number of chromosomes.

(d) *Answer:* _F_ Crossing over within the inverted region of a paracentric inversion heterozygote generates two dicentric chromosomes.

Crossing over within the inverted region of a paracentric inversion generates one dicentric chromosome and one acentric chromatid.

(e) *Answer:* _T_ An amphidiploid derived from a cross between two diploid plants is a tetraploid plant.

Amphidiploids, also called allotetraploids, are formed by doubling the number of chromosomes contained in an interspecific hybrid. Amphidiploids formed from a hybrid between two diploids are tetraploids.

(f) *Answer:* _F_ Allotetraploids are tetraploids in which the chromosome sets derive from the same species.

Allotetraploids, also called amphidiploids, are tetraploids in which the chromosome sets derived from different species.

(g) *Answer:* _F_ A child with Down syndrome (trisomy-21) has a euploid karyotype.

Euploid cells contain any multiple of the monoploid number of chromosomes (1N, 2N, 3N, etc.). Trisomy-21 (karyotype 2N+1) is an aneuploid karyotype.

(h) *Answer:* _T_ Because of their unbalanced gametes, translocation heterozygotes are usually less fertile than either translocation homozygotes or strains not containing chromosome rearrangements.

A translocation heterozygote can segregate during meiosis in two equally likely ways. "Alternate" segregation yields gametes that are balanced. "Adjacent" segregation yields gametes that are unbalanced. Thus, half of the gametes of a translocation heterozygote are unbalanced. Translocation homozygotes and strains with translocations yield only balanced gametes. Such strains are generally more fertile than translocation heterozygotes.

(i) *Answer:* _F_ An autotetraploid will have twice the number of linkage groups as a diploid of the same species.

Autotetraploids are tetraploids in which the four chromosome sets derive from the same species. Such chromosomes freely pair and cross over with each other. Although autotetraploids are often semi-sterile, they contain the same number of linkage groups as the diploids from which they derive.

(j) *Answer:* _F_ When a monosomic cell undergoes meiosis, all four of the resulting gametes are aneuploid.

In monosomic cells or organisms, one chromosome is present in one copy rather than the normal two copies (2N–1). The isolated chromosome does not have a homologue with which to pair during meiosis. The unpaired chromosome (univalent) segregates during meiosis I to one pole or the other, but not both. Half of the resulting gametes contain one copy of the chromosome and are fully haploid; the other half contain no copies of the chromosome in question and are aneuploid.

2. Fill in the blank with the defined word or phrase.

_____

(a) An alkaloid drug that blocks the assembly of spindle fibers and thus prevents the normal movement of chromosomes during mitosis or meiosis. Used to artificially produce polyploids.

_____

(b) A group of four homologous or partially homologous chromosomes that are paired with each other during meiotic prophase.

_____

(c) An inversion in which the inverted segment does not include the centromere.

_____

(d) The meiotic segregation pattern in a translocation heterozygote that results in the movement of one standard and one translocated chromosome to each of the two poles.

_____

(e) The apparent linkage between genes located on nonhomologous chromosomes involved in a translocation heterozygote. Pseudolinkage occurs because the only viable offspring produced by a translocation heterozygote result from alternate segregation, which usually carry the parental combinations of alleles.

(f) A chromosome or chromosome fragment that is lacking a centromere

_____

(g) The condition in a normally diploid cell or organism in which one chromosome is present in one copy rather than the normal two copies. Such cells or individuals have a chromosome content of 2N −1.

_____

Answer
(a) colchicine
(b) quadrivalent
(c) paracentric
(d) adjacent segregation
(e) pseudolinkage
(f) acentric
(g) monosomy

3. Seedless watermelons are triploids, containing 33 chromosomes in somatic cells. How many chromosomes are contained in the gametes of seedless watermelon?

Answer

A variable number, ranging from 11 to 22.

4. *Raphanobrassica* is an allotetraploid (amphidiploid) formed from hybrids between radish (*Raphanus sativa*) and cabbage (*Brassica oleracea*). Radishes have 9 chromosomes in their gametes. Cabbages have 18 chromosomes in their somatic cells. How many chromosomes are contained in *Raphanobrassica* gametes?

➤ *Solution*

Answer: 18

Allotetraploids are constructed by first forming hybrids between two different species, followed by induced doubling of the number of chromosomes contained in the hybrid. How many chromosomes are contained in the gametes of radishes and cabbages? The question states that radishes have 9 chromosomes in their gametes and that cabbages have 18 chromosomes in their somatic cells. Thus, both radishes and cabbages have 9 chromosomes in their gametes. A radish/cabbage hybrid contains $9 + 9 = 18$ chromosomes in its somatic cells ($2N = 18$). When this diploid number of chromosomes is doubled, the resulting allotetraploid (*Raphanobrassica*) contains 36 chromosomes in its somatic cells ($4N = 36$). The question asks how many chromosomes are contained in *Raphanobrassica* gametes? Tetraploid cells yield diploid gametes. Thus, *Raphanobrassica* contains 18 chromosomes in its gametes.

5. *Triticale* is an allohexaploid formed by crossing a tetraploid species of wheat with a diploid species of rye and allowing the hybrid to double its chromosome number. The monoploid number of both wheat and rye is 7. How many chromosomes does *Triticale* contain in its somatic cells?

Answer 42

6. The primrose *Primula kewensis* is an allotetraploid (amphidiploid) formed from a hybrid between the diploids *P. verticillata* and *P. floribunda*. *P. kewensis* contains 18 chromosomes in its gametes. *P. verticillata* and *P. floribunda* contain equal numbers of chromosomes. How many chromosomes are contained in somatic cells of *P. verticillata* and *P. floribunda*?

Answer 18

7. *Zea mays* (corn) is a diploid that contains 10 chromosomes in its gametes. How many chromosomes are contained in somatic cells of a *Z. mays* trisomic strain?

➤ *Solution*

Answer: 21

Trisomic strains are nearly diploid. They contain the normal two copies of most chromosomes, but they contain three copies of one chromosome (2N + 1). *Z. mays* contains 10 chromosomes in its gametes (N = 10). Thus, 2N = 20 and 2N + 1 = 21. A trisomic *Z. mays* strain contains 21 chromosomes in its somatic cells.

8. Meiotic cells of a wild species of *Oryza* (rice) were observed to contain 22 bivalents and 1 quadrivalent during prophase I. How many chromosomes are contained in the somatic cells of this species?

Answer 48

9. A variety of *Tripsacum dactyloides*, a wild grass, is an autotetraploid. *Zea mays* (corn) is a diploid containing 20 chromosomes in somatic cells. A hybrid is formed between *T. dactyloides* and *Z. mays*. In such hybrids, the *T. dactyloides* and *Z. mays* chromosomes can pair with homologues of their own species but not with chromosomes of the other species. Cells in meiotic prophase I of the hybrid contain 18 bivalents and 10 univalents. How many chromosomes are contained in somatic cells of the *T. dactyloides* autotetraploid?

➤ ***Solution***

Answer: 72

We can deduce the number of chromosomes in *T. dactyloides* by working backwards from the information provided. A *T. dactyloides*/*Z. mays* hybrid contains a total of 46 chromosomes (18 bivalents and 10 univalents during prophase I of meiosis). How many of these are *Z. mays* chromosomes, and how many are *T. dactyloides* chromosomes? The question states that *Z. mays* contains 20 chromosomes in somatic cells. Thus, *Z. mays* contains 10 chromosomes in its gametes. The *T. dactyloides*/*Z. mays* hybrid was formed by the fusion of one diploid *T. dactyloides* gamete and one haploid *Z. mays* gamete. Thus, 10 of the 46 chromosomes of the hybrid came from *Z. mays* and 36 came from *T. dactyloides*. If a *T. dactyloides* diploid gamete contains 36 chromosomes, then a *T. dactyloides* tetraploid somatic cell contains 72 chromosomes. Thus, *T. dactyloides* is an autotetraploid, where 4N = 72.

10. The Drosophila 4th chromosome (an autosome) is sufficiently small that both 4th chromosome monosomic and trisomic strains are viable and fertile.

(a) A trisomic male is mated with a monosomic female. Approximately what proportion of the offspring are diploid for the 4th chromosome?

(b) The *bent* gene is located on the 4th chromosome, with *bt* (bent wings) being recessive to *bt⁺* (+). A trisomic female of genotype *bt/bt/+* is crossed with a *bt/bt* diploid male. What proportion of the offspring have bent wings?

(c) The *eyeless* gene is located on the 4th chromosome, with *ey* (eyeless) being recessive to *ey⁺* (+). A trisomic male of genotype *ey/+/+* is crossed with a diploid female of genotype *ey/ey*. What proportion of the offspring are eyeless?

➤ *Solutions*

(a) *Answer:* ½

The trisomic male generates equal numbers of sperm that contain either two 4th chromosomes (diplo-4) or one 4th chromosome (haplo-4). The monosomic female generates equal numbers of eggs that contain either one 4th chromosome (haplo-4) or no 4th chromosome (nullo-4). Random unions of sperm and eggs yield a 1:2:1 ratio of trisomic:diploid:monosomic offspring. Thus, ½ of the offspring are diploid for the 4th chromosome.

(b) *Answer:* ½

To predict the genotypes of offspring, we must first predict the genotypes of the trisomic female's gametes. There are three alternative pairing configurations during meiosis of the trisomic female:

1. The two *bt* chromosomes pair as homologues, with the + chromosome segregating separately (unpaired).
2. One of the *bt* chromosomes pairs with the + chromosome, with the other *bt* chromosome segregating separately (unpaired).
3. The other *bt* chromosome pairs with the + chromosome, with the second *bt* chromosome segregating separately (unpaired).

All configurations are equally likely, and in each case, the unpaired chromosome segregates randomly during meiosis. Gametes produced by this segregation are: ⅓ *bt+* (diplo-4), ⅓ *bt* (haplo-4), ⅙ *btbt* (diplo-4), and ⅙ + (haplo-4). Only eggs of genotype *btbt* and *bt* yield offspring with bent wings. Thus, ½ of the offspring have bent wings.

(c) *Answer:* ⅙

To predict the genotypes of the offspring, we must first predict the genotypes of the male and female gametes. The female parent is a diploid of genotype *ey/ey*. Thus, all of her eggs contain a single *ey* 4th chromosome. What about the male? There are three alternative pairing configurations during meiosis of the trisomic male:

1. The two + chromosomes pair as homologues, with the *ey* chromosome segregating separately (unpaired).
2. One of the + chromosomes pairs with the *ey* chromosome, with the other + chromosome segregating separately (unpaired).
3. The other + chromosome pairs with the *ey* chromosome, with the second + chromosome segregating separately (unpaired).

All configurations are equally likely, and in each case, the unpaired chromosome segregates randomly during meiosis. Sperm produced by this segregation are: ⅓ *ey+* (diplo-4), ⅓ + (haplo-4), ⅙ ++ (diplo-4), and ⅙ *ey* (haplo-4). When used to fertilize *ey* eggs, only sperm of genotype *ey* yield eyeless offspring. Thus ⅙ of the offspring are eyeless.

---

11. An important difference between pericentric and paracentric inversions concerns the consequences of crossing over within the inverted region of inversion heterozygotes. Depending on the inversion type, crossing over within the inverted region yields unbalanced gametes either because:

1. the products of recombination are monocentric chromosomes that harbor duplications and deletions; or
2. the products of recombination are a dicentric chromosome and an acentric chromosome fragment.

Crossing over within an inversion heterozygote of Drosophila yields a dicentric chromosome and an acentric chromosome fragment. Is the inversion a pericentric or paracentric inversion?

Answer paracentric

12. Genes *A* and *B* are separated by 26 cM on chromosome 1 of Drosophila. You collect individual Drosophila from nature and isolate five different 1st chromosomes. You place these chromosomes heterozygous in pairwise combinations and measure the frequency of crossing over between the *A* and *B* genes. The results are shown in the table below.

Frequency of Recombination between Genes *A,a* and *B,b*

		CHROMOSOME ISOLATE				
		1A	1B	1C	1D	1E
	1A	~26 cM				
CHROMOSOME	1B	~6 cM	~26 cM			
ISOLATE	1C	~6 cM	~26 cM	~26 cM		
	1D	~26 cM	~6 cM	~6 cM	~26 cM	
	1E	~6 cM	~26 cM	~26 cM	~6 cM	~26 cM

If chromosome 1B is defined as "normal," which of the chromosomes contain inversion(s) relative to 1B?

Answer Chromosomes 1A and 1D

13. A normal chromosome and a paracentric inversion are shown in the figure below.

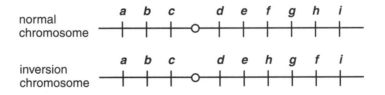

If crossing over occurs between the *f* and *g* genes in an inversion heterozygote, what genes will be contained on the acentric chromosome fragment that is produced?

Answer

One copy of genes *f*, *g*, and *h* and two copies of gene *i*.

14. A normal chromosome and a pericentric inversion are shown in the figure on p. 125.

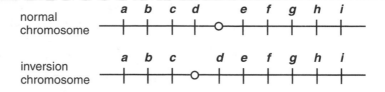

If crossing over occurs between genes *c* and *d* within the inverted region of an inversion heterozygote, what gene(s) are neither duplicated nor deleted in the two resulting recombinant chromosomes?

Answer

Gene *d*

15. Drosophila deletions that remove a particular region of the X chromosome cause a "Notch" phenotype when heterozygous. Such deletions are lethal when homozygous or hemizygous. The X-linked recessive mutation purple eyes (*pr*) is located within the region deleted by *Notch* deficiencies. Thus, *Notch* deletions cause the *pr* allele to exhibit pseudodominance. What sexual and phenotypic ratios are expected when purple-eyed Notch females are mated with normal males?

➤ **Solution**

Answer:

⅓ purple-eyed males
⅓ Notch females
⅓ normal females

To predict the phenotypes of the offspring, we must first deduce the genotypes of the parents. Let *Df(Notch)* indicate an X chromosome deletion that exhibits a Notch phenotype when heterozygous, and "+" indicate a normal X chromosome. A purple-eyed Notch female is genotype *Df(Notch)/pr*. Such females produce two types of eggs (genotypes *Df(Notch)* and *pr*). A normal male is genotype +/Y. Such males produce two types of sperm (+ and Y). Random union of eggs and sperm yield four genotypes of offspring:

$$Df(Notch)/+ = \text{Notch females (⅓ of viable offspring)}$$
$$pr/+ = \text{Normal females (⅓ of viable offspring)}$$
$$Df(Notch)/Y = \text{Lethal}$$
$$pr/Y = \text{Purple-eyed males (⅓ of viable offspring)}$$

16. The genotype of an individual heterozygous for a paracentric inversion is shown below.

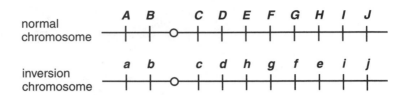

normal
chromosome

inversion
chromosome

Genes *A,a* through *J,j* are located on the chromosome and are all marked with allelic differences, such that crossovers can be detected in each interval of the chromosome. Indicate below ("Y" for yes, "N" for no) whether, among the viable offspring of an inversion heterozygote, the following single and double crossovers will be detected.

____ Single crossover in the *A,a–B,b* interval
____ Single crossover in the *C,c–D,d* interval
____ Single crossover in the *E,e–F,f* interval
____ Single crossover in the *G,g–H,h* interval
____ Single crossover in the *I,i–J,j* interval
____ Double crossover in both the *C,c–D,d* and *E,e–F,f* intervals
____ Double crossover in both the *E,e–F,f* and *G,g–H,h* intervals
____ Double crossover in both the *C,c–D,d* and *I,i–J,j* intervals

Answer

_Y_ Single crossover in the *A,a–B,b* interval.
_Y_ Single crossover in the *C,c–D,d* interval
_N_ Single crossover in the *E,e–F,f* interval
_N_ Single crossover in the *G,g–H,h* interval
_Y_ Single crossover in the *I,i–J,j* interval
_N_ Double crossover in both the *C,c–D,d* and *E,e–F,f* intervals
_Y_ Double crossover in both the *E,e– F,f* and *G,g–H,h* intervals
_Y_ Double crossover in both the *C,c– D,d* and *I,i–J,j* intervals

17. The relative structures of two chromosomes from related species are shown in the diagram on p. 127.

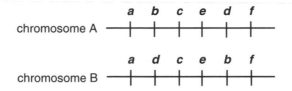

What is the minimum number of inversions needed to derive chromosome B from chromosome A?

➤ *Solution*

Answer: Two

The relationships of genes on chromosomes change during evolution by chromosome rearrangement. How are chromosomes A and B different? The relative positions of genes *b* and *d* are exchanged. How can this be accomplished by a series of inversions? First, if a segment of chromosome A including genes *b*, *c*, *e*, and *d* is inverted, the resulting chromosome is structure *a-d-e-c-b-f*. This places genes *a*, *d*, *b*, and *f* in the same relative positions as those found in chromosome B, but genes *e* and *c* are in the incorrect order. A second inversion, inverting the segment of chromosome containing *e* and *c*, can then generate chromosome B (*a-d-c-e-b-f*).

18. You isolate 3 lethal mutations of Drosophila following X-ray mutagenesis. Each of these mutations is wild-type when heterozygous and lethal when homozygous. Cytological analysis of salivary gland polytene chromosomes demonstrates that the three mutations, *Df1*, *Df2*, and *Df3*, are small deletions (deficiencies). Five single-gene lethal mutations, *let1* through *let5*, are located near the deletions, but their exact order is unknown. You cross each deletion with each single-gene lethal and determine whether the resulting heterozygote is viable (+ in the table below) or lethal (− in the table below).

Viability of *Df*/*let* Heterozygotes

	let1	*let2*	*let3*	*let4*	*let5*
Df1	+	−	−	−	+
Df2	−	−	+	+	+
Df3	−	−	+	−	−

+ = viable − = lethal

Draw the linkage map of the region, indicating the relative positions of *Df1*, *Df2*, *Df3*, and *let1* through *let5*.

➤ Solution

Answer:

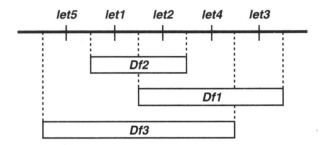

The key to constructing the linkage map is understanding that whenever a *Df/let* heterozygote is lethal, the deletion removes the *let* gene, and whenever a *Df/let* heterozygote is viable, the deletion does not remove the *let* gene. Deletions remove a contiguous region of the chromosome. Thus, we must figure out what linear order of *let* genes is consistent with each of the deletions removing a contiguous region of material.

We can work our way piecemeal through the data. For example, *Df1* deletes *let2*, *let3*, and *let4*, but it does not delete *let1* and *let5*. We can begin to construct the linkage map as follows:

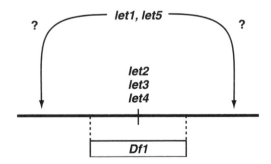

This incomplete map indicates that *Df1* deletes *let2*, *let3*, and *let4*, and that *let1* and *let5* must map either to the left or to the right of *Df1*. Now, we can add information for *Df2*. *Df2* deletes *let1* and *let2*, but does not delete *let3*, *let4*, or *let5*. What order of genes is consistent with this information? If we assume that the deletions remove a contiguous region of the chromosome, *let1* must be located to the left of *let2*, and *let2* must be located to the left of both *let3* and *let4*. The linkage map can, therefore, be refined as follows:

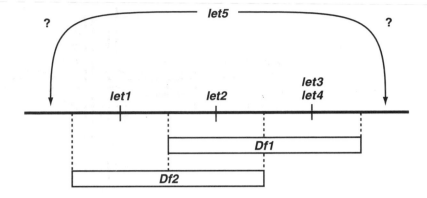

This refined map indicates the genes that are deleted by *Df1* and *Df2*. Left unresolved, however, are (1) whether *let5* maps to the left of *Df2* or to the right of *Df1* and (2) the relative order of *let3* and *let4*. These uncertainties are clarified by adding the information for *Df3*. *Df3* deletes all genes except *let3*. What order of genes is consistent with this information? There is only one order of *let* genes that allows each deletion to remove a contiguous region of material. *Let3* must be located to the right of *let4*, and *let5* must be located to the left of *let1*. The linkage map, therefore, can be refined into an unambiguous order:

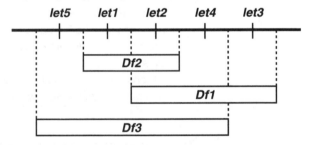

Note that the reverse order (*let3-let4-let2-let1-let5*, left-to-right) is equally correct.

19. The diagram on p. 130 shows a translocation heterozygote paired at meiosis.

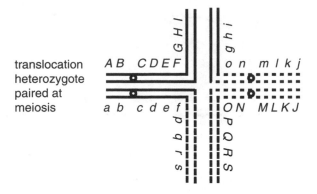

translation
heterozygote
paired at
meiosis

What proportion of the gametes obtained from this translocation heterozygote are unbalanced, containing deletions of some sequences and duplications of others?

> ### Solution
>
> *Answer:* 50%
>
> There are two equally likely ways in which a translocation heterozygote can segregate during meiosis. In "alternate" segregation, the two unre-arranged chromosomes go to one pole and the two halves of the translocation go to the other pole during meiosis I. Gametes produced by alternate segregation are balanced, containing a full complement of all genes from both chromosomes. In "adjacent" segregation, one unre-arranged chromosome and one half of the translocation go to each pole during meiosis I. Such gametes are unbalanced, containing dupli-cations and deletions of chromosomal material. Thus, half the gametes of a translocation heterozygote are balanced, and half are unbalanced.

---

20. You identify seven recessive lethal mutations of the nema-tode, *m1* through *m7*, that map within a small region of the linkage map. Each of these mutations is wild-type when heterozygous and lethal when homozygous. You cross these seven mutations in all pairwise combinations and deter-mine whether the resulting heterozygotes (*m/m*) are viable (+ in the table on p. 131) or lethal (– in the table on p. 131).

Viability of *m/m* Heterozygotes

	m1	m2	m3	m4	m5	m6	m7
m1	–						
m2	+	–					
m3	–	+	–				
m4	+	+	+	–			
m5	+	+	+	+	–		
m6	+	–	+	+	–	–	
m7	+	+	+	–	+	+	–

+ = viable – = lethal

Which of these mutations is likely to be a deletion?

Answer Mutation *m6*

21. Genes *A* and *B*, found in your favorite plant, are unlinked. Thus, if you cross a wild-type plant (genotype *AB*) with a recessive homozygote (genotype *ab*), and subsequently test-cross the F1 dihybrid (genotype *A/a B/b*) to the recessive homozygote, genes *A* and *B* exhibit independent assortment (1:1:1:1 ratio of phenotypes *AB*, *Ab*, *aB*, and *ab*). You have found a wild-type stock, however, that gives unexpected results. When you cross this unusual wild-type stock (genotype *AB*) with the recessive homozygote, and then testcross the F1 dihybrid to the recessive homozygote, genes *A* and *B* appear to be linked. Most offspring are phenotype *AB* or *ab*, few are phenotype *Ab* or *aB*.

(a) Assuming that genes *A* and *B* are located on different chromosomes in normal plants, what chromosomal rearrangement(s) in the exceptional wild-type stock would explain these results?

(b) Assuming that genes *A* and *B* are located on the same chromosome in normal plants, what chromosomal rearrangement(s) in the exceptional wild-type stock would explain these results?

➤ *Solutions*

(a) *Answer:* A reciprocal translocation

A reciprocal translocation involving the chromosomes on which genes *A* and *B* reside can cause them to appear linked. The unusual wild-type is probably a translocation homozygote. When this wild-type strain is crossed with the *ab* double mutant, the resulting F1 dihybrid is a translocation heterozygote. Genes located within the translocated segments exhibit pseudolinkage when segregating from translocation heterozygotes.

(b) *Answer:* An inversion

If genes *A* and *B* are located on the same chromosome of normal plants yet are unlinked, they must be sufficiently far apart that frequent crossing over between them causes them to assort independently. What chromosomal rearrangement of this chromosome will decrease the apparent frequency of recombination between *A* and *B*? An inversion that inverts much of the chromosome between *A* and *B* would cause them to appear linked. The inversion could include either the *A* or *B* genes (or both), but the important aspect of the inversion is that it should invert the region *between* genes *A* and *B*. Crossing over within the inverted region of a pericentric or paracentric inversion causes the resulting gametes to be unbalanced and, hence, not recovered among viable offspring. If the exceptional wild-type plant were homozygous for such an inversion, the F1 dihybrid would be an inversion heterozygote, in which case crossing over between genes *A* and *B* would be reduced or eliminated.

22. Two chromosomes pair in meiosis to produce the following configuration.

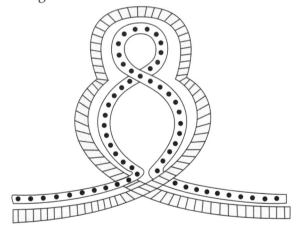

Describe the chromosomal rearrangement(s) that one of the chromosomes contains.

Answer

One of the chromosomes contains a small inversion contained completely within a larger inversion.

23. You have isolated a collection of X-linked, recessive, lethal mutations of Drosophila. Such mutations are recessive to wild-type, but lethal when homozygous in females or hemizygous in males. You examine the polytene chromosomes of your mutants and recognize that four of them contain deletions (deficiencies) of defined portions of the X chromosome. You name these deletion mutations *Df1*, *Df2*, *Df3*, and *Df4*.

In your collection of Drosophila strains, you have nine X-linked recessive mutations that affect various aspects of the fly's phenotype (wing shape, eye color, bristle morphology, etc.). You know that these mutations are located on the X chromosome because they show sex-linked inheritance. These nine mutations are named *v1*, *v2*, *v3*, *v4*, etc., where "*v*" stands for "visible phenotype."

You cross each of the four deletion-containing strains with each of the nine visible-containing strains and test whether the deletions "uncover" the visible alleles; that is, whether the deletions cause the visibles to be pseudodominant. You obtain the following data:

					VISIBLE MUTATION				
Df	**v1**	**v2**	**v3**	**v4**	**v5**	**v6**	**v7**	**v8**	**v9**
Df1	1	2	+	4	+	6	+	8	9
Df2	+	2	+	+	+	+	7	8	9
Df3	1	+	3	4	5	+	+	+	+
Df4	1	+	+	4	5	6	+	8	+

In each case, a "+" indicates that the *Df/v* heterozygote is normal. A number indicates that the heterozygote exhibits the phenotype of the visible allele.

(a) Draw a linkage map of the chromosome, showing the order of each of the visible alleles and the regions deleted by each of the deficiencies.

(b) You notice that mutants *v3* and *v5* both have the same visible phenotype (curled wings). Is it likely that these two mutations affect the same gene?

(c) You notice that, although mutants *v2* and *v9* have slightly differing phenotypes (pink eyes and carnation eyes), they are both located in the same region of the X chromosome. Is it possible that these two mutations affect the same gene?

(d) Predict the phenotype of a female whose genotype is *Df2/Df3*.

Answer

(a)

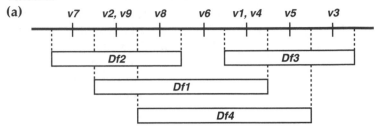

(b) No, it is not likely.
(c) Yes, it is possible.
(d) Wild type

 24. Genes encoding lanosterol synthase (*LSS*) and the liver form of phosphofructokinase (*PFKL*) are located on human chromosome 21. A mother and father of normal karyotype have a daughter with Down syndrome (trisomy-21). The inheritance of *LSS* and *PFKL* alleles in this family is shown in the diagram below.

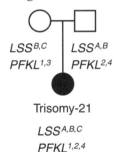

$LSS^{B,C}$ $LSS^{A,B}$
$PFKL^{1,3}$ $PFKL^{2,4}$

Trisomy-21

$LSS^{A,B,C}$
$PFKL^{1,2,4}$

In which parent and at which stage of meiosis did nondisjunction occur, yielding the trisomy-21 child? (Ignore the consequences of crossing over in either parent when answering this question.)

➤ **Solution**

Answer: Meiosis I of the father

We can deduce where nondisjunction occurred by examining inheritance of *LSS* and *PFKL* alleles. Look first at inheritance of *LSS*. The trisomy-21 daughter is genotype $LSS^{A,B,C}$. Two of these alleles were inherited from one parent and one was inherited from the other parent. But which two alleles came from one parent? Based on the *LSS* genotype, we cannot be certain. A child of genotype $LSS^{A,B,C}$ might have arisen in two different ways:

1. Nondisjunction occurs in the mother; she contributes alleles LSS^B and LSS^C. The father contributes allele LSS^A.

2. Nondisjunction occurs in the father; he contributes alleles LSS^A and LSS^B. The mother contributes allele LSS^C.

Thus, inheritance of LSS alleles leaves ambiguous the origins of the nondisjunction. What about inheritance of $PFKL$ alleles?

The trisomy-21 daughter is genotype $PFKL^{1,2,4}$. Two of these alleles were inherited from one parent and one was inherited from the other parent. In this case, however, it is unambiguous as to which parent contributed two alleles. The child contains alleles $PFKL^2$ and $PFKL^4$, both of which are contained by the father and neither of which are contained by the mother. Thus, the child inherited two $PFKL$ alleles from her father and one ($PFKL^1$) from her mother.

In which meiotic division did nondisjunction occur in the father? Ignoring the possible consequences of crossing over in the father, nondisjunction occurring during meiosis II of the father would generate gametes containing either two copies of the $PFKL^2$ chromosome or two copies of the $PFKL^4$ chromosome. Nondisjunction occurring during meiosis I of the father would generate gametes containing one copy of both the $PFKL^2$ and $PFKL^4$ chromosomes. In the absence of crossing over, only nondisjunction occurring during meiosis I of the father is consistent with the observed pattern of inheritance.

---

25. The most common form of polyploidy in humans is triploidy. Human triploids are inviable, dying as young embryos *in utero*. Approximately 17% of spontaneous abortions are caused by triploid fetuses. Most triploids result from dispermy, the simultaneous fertilization of a haploid egg by two haploid sperm. Such triploids contain one maternal and two paternal sets of chromosomes. The human monoploid number is 23. Which of the following triploid karyotypes is the most common: 69, XXX; 69, XXY; or 69, XYY?

➤ *Solution*

Answer: 69, XXY

Triploids resulting from dispermy are fertilized by two separate haploid sperm. If both of those sperm contain X chromosomes, the triploid is 69, XXX. How often do two sperm both contain X chromosomes? The probability of any single sperm containing an X chromosome is ½. Thus (½)(½) = ¼ of pairs of sperm both contain X chromosomes. Similarly, ¼ of pairs of sperm both contain Y chromosomes. How about one X and one Y chromosome? Regardless of whether the *first* sperm contains an X or a Y chromosome, the probability that the *second* sperm will contain a sex chromosome *different* from the first is ½. Thus, approximately ¼ of triploid fetuses are 69, XXX, ¼ are 69, XYY, and ½ are 69, XXY.

Bacterial
and Phage Genetics

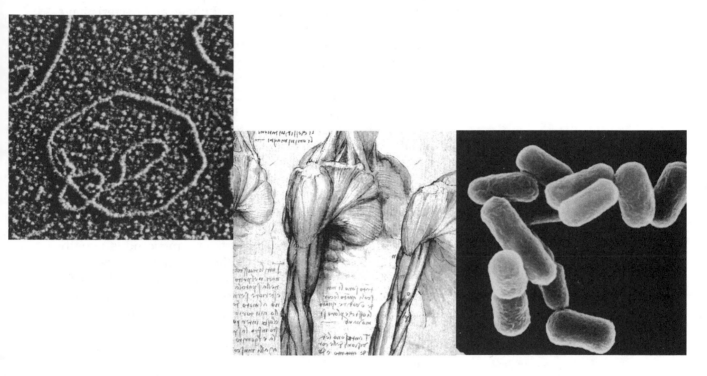

Summary

Bacteria are **prokaryotes**, small unicellular organisms whose single circular chromosome is not surrounded by a nuclear envelope. The life cycle of prokaryotes is, in many respects, simpler than that of organisms containing nuclei (**eukaryotes**). Prokaryotes are almost always haploid, they do not undergo meiosis or form gametes, and the mechanics of prokaryotic cell division are visibly less complex than those of eukaryotes. **Bacteriophage** (more commonly called **phage**) are viruses that infect bacteria. Genetic investigations of phage and bacteria have been instrumental to our understanding of gene structure, gene function, and the molecular mechanisms that underlie heredity. Genetic manipulation of the common intestinal bacterium *Escherichia coli* is particularly significant because it provides many of the tools of recombinant DNA technology and genetic engineering.

Genetic exchange in bacteria differs from that of multicellular eukaryotes in several important ways:

1. Bacteria do not make gametes. Rather, genetic material (DNA) is directly transferred from one bacterial cell to another.
2. Transmission of DNA is unidirectional. Individual cells are either **donors** or **recipients** (but not both) of such exchange.
3. Only a small fraction of a donor's chromosome is usually transmitted to a recipient during the exchange process.
4. DNA received by a recipient cell is not replicated and inherited unless it first recombines with the recipient's circular chromosome, thereby replacing that region of the chromosome.

Bacteria participate in three different forms of gene transfer. These forms of exchange differ from each other by the manner in which segments of the chromosome are transferred from one bacterial cell to another. During **transformation**, bacteria directly transport DNA from the growth medium into the cell. During **conjugation**, specialized properties of the **fertility factor** cause DNA to be transferred from one cell to a nearby cell with which it is physically connected. During **generalized transduction**, viral particles produced during a cycle of phage infection contain bacterial chromosomal DNA instead of phage DNA. Such phage particles transport DNAs from donor to recipient cells.

Many species of bacteria actively transport DNA from the growth medium into the cell. Other species, such as *E. coli,* do not naturally transport DNA, but they can be induced to do so under special conditions. For those species that naturally transport DNA, only a single strand of the initially double-stranded DNA is transported into the cell. After entering a recipient cell, such DNA can recombine with the recipient's circular chromosome. The process by which cells receive and inherit genetic markers following direct uptake of DNA from the medium is termed transformation, and cells whose phenotype is altered by transformation are termed **transformants**. DNAs transferred in this manner are usually small linear fragments of the donor's large circular chromosome. If two donor alleles are sufficiently close together that they are located on a single DNA molecule, they can be simultaneously inherited by a single transformation event. The frequency of such **cotransformation** is a measure of the physical distance between the alleles. Alleles located close together exhibit a high frequency of cotransformation; those farther apart exhibit low frequencies of cotransformation.

Specialized properties of the fertility factor, F, cause bacteria to transfer DNA by conjugation. Strains containing a fertility factor (F$^+$, F′, or Hfr strains) transmit genes only to those not containing one (F$^-$ strains). In F$^+$ and F′ bacteria, the fertility factor is inherited as a **plasmid**, a small circular DNA molecule separate from the larger bacterial chromosome. F′ plasmids differ from F$^+$ plasmids in that F′ plasmids contain some bacterial genes attached to the fertility factor. During conjugation, DNA replication produces a single strand of the F$^+$ or F′ plasmid, which is then transferred to an F$^-$ recipient. A complementary DNA strand is synthesized in the recipient. After transfer of a complete F$^+$ or F′ genome, a circular F$^+$ or F′ plasmid is reformed in the recipient. In **Hfr** strains, a fertility factor is integrated into the bacterial chromosome. Because of the transfer properties of the fertility factor, Hfr strains efficiently transmit chromosomal genes to F$^-$ recipients. Hfr transfer of chromosomal genes begins at a specific point within the fertility factor and proceeds in one direction around the donor chromosome. Depending on the site of fertility factor integration, different Hfr strains have different origins and directions of gene transfer. A single DNA strand, produced by DNA replication, is transmitted during Hfr transfer, with the complementary strand being synthesized in the recipient. Hfr strains can, in principle, transfer their entire chromosome to an F$^-$ recipient, a process that requires approximately ninety minutes to complete. The order of gene transfer during an Hfr mating can be deduced by determining the earliest times at which genes are inherited during an **interrupted mating** experiment.

Alternatively, the order of transferred genes can be deduced by measuring the relative frequencies with which those genes are inherited during an uninterrupted mating experiment. Mating pairs spontaneously break apart during conjugation, even without the deliberate interruption of mating. This causes genes transferred early by an Hfr to be inherited at a higher frequency than those transferred late.

During growth of certain bacteriophage, a small fraction of assembled phage heads contains DNA of the host bacterium rather than DNA of the phage. Such phage particles transfer bacterial genes to cells that they subsequently infect, a process termed **generalized transduction**. An individual transducing phage particle contains only a small fragment of the bacterial chromosome (about 1% of its length), but a large collection of transducing phages collectively contains the entire donor bacterial chromosome. When such phages infect bacteria, some infected cells receive bacterial, rather than phage, DNA. Transduced DNA can then recombine with the chromosome of the recipient cell. Alleles located close together on the bacterial chromosome are often contained on a single transduced DNA fragment. Such alleles can be simultaneously transduced (**cotransduced**) following infection of a recipient with a single transducing phage particle. The **frequency of cotransduction**, defined as the proportion of transductants for one allele that simultaneously inherits a nearby allele, is a measure of the physical distance between those alleles. Alleles located close together have high frequencies of cotransduction; alleles located far apart have low frequencies of cotransduction.

Like all viruses, bacteriophage are intracellular parasites. During **lytic** cycles of growth, infecting phages usurp the host's biosynthetic machinery and reproduce tens or hundreds of new virus particles, eventually breaking open the host cell. **Temperate** bacteriophage can follow either the lytic or the **lysogenic** life cycle. During the lysogenic cycle, an infecting phage does not kill the host cell. Rather, the virus enters a quiescent state, during which most phage gene expression is shut off. The phage genome integrates into the chromosome of its bacterial host and replicates as a part of it. A bacterium containing such a quiescent phage is termed a **lysogen**, and the integrated phage genome is termed a **prophage**. Certain environmental stimuli induce prophages to excise from the host chromosome and enter the lytic life cycle.

Genetic crosses between bacteriophage are performed by simultaneously infecting bacteria with two different phage mutants and allowing a single cycle of lytic phage growth. Recombinants are then scored among the progeny phage. The recombination frequency (RF) between two alleles is measured as the proportion of recombinant genotypes among total phage progeny. Alleles located close together exhibit a low RF; alleles located far apart exhibit a high RF. Fine structure genetic maps are constructed from such RF measurements. Phage complementation tests determine whether two mutations that block phage growth affect the same or different genes. Such tests are performed by simultaneously infecting bacteria with two different phage mutants each of which is individually unable to complete the lytic life cycle. If the two mutations affect *different* genes, then coinfected cells yield normal numbers of progeny phage. The progeny phage in this case are parental genotypes plus a few recombinants. If the two mutations affect the *same* gene, then coinfected cells yield no progeny phage except for a few recombinants.

Self-Testing Questions

1. Are the following statements true or false?

____   **(a)** A tryptophan prototroph cannot grow unless tryptophan is added to the growth medium.

____   **(b)** Two genes that exhibit a high frequency of cotransduction are closer together than two genes that exhibit a low frequency of cotransduction.

____   **(c)** A fertility factor can be transmitted to a recipient cell only as part of an Hfr × F⁻ cross.

____   **(d)** During an Hfr cross, the very first DNA to enter a recipient cell is DNA from the fertility factor itself.

____   **(e)** An F′ element always contains some bacterial chromosomal DNA.

____   **(f)** Strains containing F′ elements are merodiploid cells.

____   **(g)** If 10^5 phage are used to infect 10^6 bacteria, and if these infected bacteria produce 10^7 progeny phage, the burst size for this infection is 100.

____   **(h)** If two million phage particles are used to infect one million bacterial cells, the multiplicity of infection is 0.5.

____   **(i)** If two mutations fail to complement, then they affect the same gene.

____   **(j)** A bacterial lysogen contains one or more prophages as part of its genome.

_____   **(k)** An Hfr cell will not conjugate with an F$^+$ cell.

_____   **(l)** A generalized transducing phage must lysogenize its host before it can transduce chromosomal genes.

_____   **(m)** In Hfr strains, the fertility factor is inherited as a plasmid.

Answer

_F_   **(a)** A tryptophan prototroph cannot grow unless tryptophan is added to the growth medium.

_T_   **(b)** Two genes that exhibit a high frequency of cotransduction are closer together than two genes that exhibit a low frequency of cotransduction.

_F_   **(c)** A fertility factor can be transmitted to a recipient cell only as part of an Hfr × F$^-$ cross.

_T_   **(d)** During an Hfr cross, the very first DNA to enter a recipient cell is DNA from the fertility factor itself.

_T_   **(e)** An F′ element always contains some bacterial chromosomal DNA.

_T_   **(f)** Strains containing F′ elements are merodiploid cells.

_T_   **(g)** If 10^5 phage are used to infect 10^6 bacteria, and if these infected bacteria produce 10^7 progeny phage, the burst size for this infection is 100.

_F_   **(h)** If two million phage particles are used to infect one million bacterial cells, the multiplicity of infection is 0.5.

_T_   **(i)** If two mutations fail to complement, then they affect the same gene.

_T_   **(j)** A bacterial lysogen contains one or more prophages as part of its genome.

_T_   **(k)** An Hfr cell will not conjugate with an F$^+$ cell.

_F_   **(l)** A generalized transducing phage must lysogenize its host before it can transduce chromosomal genes.

_F_   **(m)** In Hfr strains, the fertility factor is inherited as a plasmid.

2. Fill in the blank with the defined word or phrase.

_____   **(a)** The process by which DNA is transferred from a donor cell that contains a fertility factor to a recipient cell that does not contain a fertility factor through a close union of the two cells.

_____   **(b)** One of the two reproductive pathways available to a temperate bacteriophage upon infecting a host cell. In this pathway, the phage chromosome establishes a state of repression in which most phage genes are not expressed and progeny phage are not produced. The phage DNA usually integrates into the bacterial chromosome in this pathway.

_____ (c) A nutritional mutant in bacteria or fungi that can only grow on an appropriately supplemented medium.

_____ (d) The process by which infecting bacteriophages package DNA of the host bacterial chromosome and transmit those chromosomal genes to subsequent infected cells.

_____ (e) The number of progeny phage particles released upon lysis of an infected host cell.

_____ (f) The DNA of a temperate bacteriophage that has established a state of repression within its bacterial host. Most phage genes are not expressed, progeny phage are not produced, and phage DNA is replicated as a part of host cell division.

_____ (g) The process by which competent bacterial cells transport DNA from the growth medium into the cell and incorporate it as part of their genetic makeup.

Answers
(a) conjugation
(b) lysogeny
(c) auxotroph
(d) generalized transduction
(e) burst size
(f) prophage
(g) transformation

3. You are given an *E. coli* strain that is a triple mutant. This strain contains: 1. a heat-sensitive mutation that causes histidine auxotrophy; 2. a cold-sensitive mutation that causes arginine auxotrophy; and 3. a non-conditional mutation that causes streptomycin resistance. Predict the growth properties of this strain on the following media. For the heat-sensitive mutation, 30° is the permissive temperature and 37° is the non-permissive temperature. For the cold-sensitive mutation, 37° is the permissive temperature and 30° is the non-permissive temperature.

His = histidine
Arg = arginine
Str = streptomycin

MINIMAL		MINIMAL+HIS		MINIMAL+ARG		MINIMAL+HIS+ARG		MINIMAL+ARG+STR	
30°	37°	30°	37°	30°	37°	30°	37°	30°	37°
___	___	___	___	___	___	___	___	___	___

(Fill in a "+" or a "−" to indicate whether the strain will grow.)

Answer

MINIMAL		MINIMAL+HIS		MINIMAL+ARG		MINIMAL+HIS+ARG		MINIMAL+ARG+STR	
30°	37°	30°	37°	30°	37°	30°	37°	30°	37°
−	−	−	+	+	−	+	+	+	−

4. You are given an *E. coli* strain of unknown genotype. It contains one or more mutations. You perform the following growth tests.

MIN		MIN+TRP		MIN+LEU		MIN+TRP+LEU		MIN+TRP+ LEU+AMP	
30°	37°	30°	37°	30°	37°	30°	37°	30°	37°
−	−	+	−	−	−	+	+	−	+

> \+ = the strain grows under the indicated conditions
> − = the strain does not grow under the indicated conditions
> Min = minimal medium
> Trp = tryptophan, an amino acid
> Leu = leucine, an amino acid
> Amp = ampicillin, an antibiotic

(a) List the mutations that this strain contains and describe their effects on the phenotype of the strain. Are any of the mutations conditional? If so, indicate how temperature affects each of the mutant phenotypes.

(b) Predict whether the strain will grow on the following media:

MIN+TRP+AMP		MIN+LEU+AMP	
30°	37°	30°	37°
___	___	___	___

Answer

(a) The strain contains three mutations:

1. a *trp⁻* mutation that is not conditional (*trp⁻* at both 30° and 37°)

2. a temperature sensitive *leu⁻* mutation (*leu⁺* at 30°, *leu⁻* at 37°)

3. a temperature sensitive ampicillin resistance mutation (Amp^S at 30°, Amp^R at 37°)

(b)

MIN+TRP+AMP		MIN+LEU+AMP	
30°	37°	30°	37°
–	–	–	–

5. You are given a strain of *E. coli* that contains one or more mutations. You spread 10^8 cells of this strain onto eight different Petri dishes that contain the media indicated below.

  Min = Minimal media
  +His = the amino acid histidine is added
  +Trp = the amino acid tryptophan is added
  +Tet = the antibiotic tetracycline is added. *E. coli* is normally sensitive to the killing effects of tetracycline (Tet^S), but resistant mutants (Tet^R) are known.

The following patterns of growth occur on the eight plates:

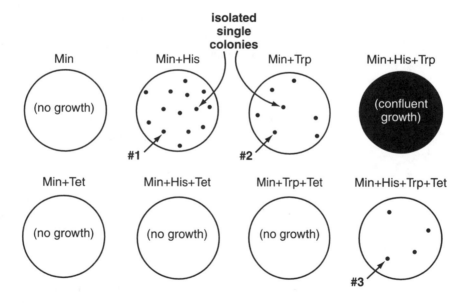

(a) What is the genotype of this strain (describe any amino acid auxotrophy or drug resistance)?

(b) You isolate single colonies #1, #2, and #3 and retest their ability to grow on the following media. Indicate with a "+" or "–" whether cells from the isolated colony will grow.

+ = Cells will grow
− = Cells will not grow

	Min	Min+His	Min+Trp	Min+His+Trp	Min+His+ Trp+Tet
Colony #1	——	——	——	——	——
Colony #2	——	——	——	——	——
Colony #3	——	——	——	——	——

➤ *Solutions*

(a) *Answer:* The strain is $his^-trp^-tet^S$.

We must deduce the genotype of the strain from its growth proper-ties. Of the eight tested media, the strain of unknown genotype grows on only one of them (Min+His+Trp). By comparing its growth properties on Min+His+Trp to those on other media, we can decide individually whether the strain is His$^+$ or His$^-$, Trp$^+$ or Trp$^-$, and Tet^S or Tet^R. For example, the strain grows on Min+His+Trp, but it does not grow on Min+His+Trp+Tet. Thus, the strain is sensitive to tetra-cycline. Similarly, the strain grows on Min+His+Trp, but it does not grow on Min+Trp. Thus, the strain is His$^-$. Such comparisons indi-cate that the strain is genotype $his^-trp^-tet^S$.

It is important to not confuse growth of the isolated single colonies with growth of the starting strain. One hundred million (10^8) cells of the starting strain were placed onto each Petri dish. If these cells are capable of growth on the medium, a confluent lawn of cells would result (e.g., Min+His+Trp). In cases where a small number of colonies grow, each colony represents the clonal descendants of a single cell whose genotype is different than that of the vast majority of tested cells.

(b) *Answer:*

	Min	Min+His	Min+Trp	Min+His+Trp	Min+His+ Trp+Tet
Colony #1	−	+	−	+	−
Colony #2	−	−	+	+	−
Colony #3	−	−	−	+	+

In part **(a)**, we determined the starting strain to be genotype his^-trp^- tet^S. Single colony #1 grows on Min+His medium. Thus, it is a Trp$^+$ revertant. Its genotype is his^-tet^S. Colony #1 will only grow if histi-dine is supplied in the medium and tetracycline is absent. Single colony #2 grows on Min+Trp medium. Thus, it is a His$^+$ revertant. Its genotype is trp^-tet^S. Colony #2 will only grow if tryptophan is sup-plied in the medium and tetracycline is absent. Colony #3 grows on Min+His+Trp+Tet. Thus, it is a tetracycline resistant (Tet^R) mutant.

Its genotype is *tet^R*. Colony #3 will grow whenever both histidine and tryptophan are supplied in the medium.

6. Bacterial DNA harvested for use in transformation experiments is unavoidably sheared by mechanical stress into small fragments, usually about 15,000 base pairs average length. Two different samples of wild-type *Streptococcus* DNA are used to transform an *A^-B^-* recipient strain. Using DNA sample #1, genes *A* and *B* exhibit 60% cotransformation. Using DNA sample #2, genes *A* and *B* exhibit 35% cotransformation. In which DNA sample is the average length of the wild-type DNA fragments larger?

Answer

The average length of DNA fragments in sample #1 is larger than that of sample #2.

7. Mutations affecting the *his*, *trp*, and *tyr* genes of *Bacillus subtilis* cause auxotrophy of histidine, tryptophan, and tyrosine, respectively. All three genes are located close together. A *his^- trp^- tyr^-* recipient is transformed with DNA from a *his^+ trp^+ tyr^+* donor. The number of colonies of each transformant class is shown in the table below.

Class	Phenotype	Number of Colonies
1	His^+Trp^+Tyr^+	11,940
2	His^+Trp^-Tyr^+	3,660
3	His^-Trp^+Tyr^-	2,600
4	His^+Trp^+Tyr^-	1,180
5	His^-Trp^-Tyr^+	685
6	His^+Trp^-Tyr^-	418
7	His^-Trp^+Tyr^+	107

What is the relative order of *his*, *trp*, and *tyr* on the *B. subtilis* linkage map?

➤ **Solution**

Answer: trp-his-tyr (or the reverse order, which is equivalent)

The fact that His^+Trp^+Tyr^+ transformants can be obtained at high frequency demonstrates that all three genes are closely linked. Thus, *his^+*, *trp^+*, and *tyr^+* can all be inherited on a single molecule of transforming DNA. But what is the order of genes on this DNA? We can deduce the relative order of these three genes from the number of transformants obtained of each genotype.

Transforming DNA is linear, and the recipient chromosome into which it recombines is circular. Thus, an even number of reciprocal crossovers

(2, 4, etc.) is required to maintain the circularity of the recipient chromosome. Two crossovers substitute a contiguous region of the recipient's chromosome with the corresponding region of donor DNA. Donor genes inherited in this way must be contiguous. Four crossovers, however, substitute two nearby but non-contiguous regions of the recipient chromosome with donor material. If, for example, donor DNA contains the genes *A-B-C* (in that order), then inheritance of *A* and *C* (but not *B*) can occur only by a quadruple crossover. It cannot occur by a double crossover. Quadruple crossovers are less frequent than any double crossovers. Which recombinants are the least frequent? In the above table, the least frequent class of recombinant is the His⁻Trp⁺Tyr⁺ class. From the cross *his⁺trp⁺tyr⁺* × *his⁻trp⁻tyr⁻*, which order of genes causes His⁻Trp⁺Tyr⁺ recombinants to result from quadruple crossovers? Only one order, *trp-his-tyr*, has this effect.

8. The *argB* gene of *B. subtilis* is unlinked to both *trpA* and *trpB*. An *argB⁻ trpA⁻* recipient is transformed with DNA either from wild-type (genotype *argB⁺trpA⁺trpB⁺*) or from a *trpB⁻* mutant (genotype *argB⁺trpA⁺trpB⁻*). DNA-treated cells are divided and the number of Arg⁺ and Trp⁺ transformants are determined separately. Results are shown in the table below.

Donor DNA	Number of Arg⁺ Transformants	Number of Trp⁺ Transformants
argB⁺trpA⁺trpB⁺	1,000	1,000
argB⁺trpA⁺trpB⁻	1,000	350

What is the frequency of cotransformation of *trpA* and *trpB*?

➤ *Solution*

Answer: 65%

With *trpA⁺trpB⁺* donor DNA, 1,000 Trp⁺ transformants are obtained. With *trpA⁺trpB⁻* donor DNA, only 350 Trp⁺ transformants are obtained. Why are there fewer Trp⁺ transformants with *trpB⁻* donor DNA? Where are the "missing" Trp⁺ transformants? They are not really missing; they are simply Trp⁻ due to the fact that they inherited the donor's *trpB⁻* allele along with a *trpA⁺* allele.

If *trpA* and *trpB* are close together, they can be cotransformed with the same molecule of DNA. Using *trpA⁺trpB⁺* donor DNA, all recipients that inherit *trpA⁺* are recovered as Trp⁺ transformants. Using *trpA⁺trpB⁻* donor DNA, however, Trp⁺ transformants are recovered only when the recipient inherits *trpA⁺* but does *not* inherit *trpB⁻*. Thus, when *trpA⁻* recipients are treated with *trpA⁺trpB⁻* DNA, 350 transformants are *trpA⁺trpB⁺* and 650 are *trpA⁺trpB⁻*. Thus, the percent cotransformation between *trpA* and *trpB* is 65%.

9. Four different *E. coli* Hfr strains donate the following markers in the order indicated below:

Hfr #1: (first)...*Q*...*W*...*D*...*M*...*T*...
Hfr #2: (first)...*A*...*X*...*P*...*T*...*M*...
Hfr #3: (first)...*B*...*N*...*C*...*A*...*X*...
Hfr #4: (first)...*B*...*Q*...*W*...*D*...*M*...

All of these Hfr strains are derived from the same F⁺ strain. Draw a linkage map of the circular bacterial chromosome showing the relative position of each marker and the origins and directions of transfer of Hfrs #1–4.

Answer

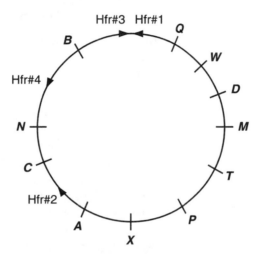

10. The earliest marker transferred by Hfr C is *purE⁺*. Among *purE⁺* conjugants selected from the cross Hfr C *purE⁺ argE⁻ lac⁺ thr⁺ proA⁺* × F⁻ *purE⁻ argE⁺ lac⁻ thr⁻ proA⁻*, 90% are Lac⁺, 82% are ArgE⁺, 65% are ProA⁺, and 33% are Thr⁺. What is the order of transfer of these five genes by Hfr C?

Answer
purE is transferred first, followed by *lac, proA, thr,* and *argE* (in that order).

11. *E. coli* strains A1, A2, A3, and A4 are *arg⁻ leu⁺*. Strains L1, L2, L3, and L4 are *arg⁺ leu⁻*. Some of these strains are Hfr, some are F⁺, and some are F⁻. Strains A1 through A4 are individually mixed with strains L1 through L4 and placed on minimal medium, selecting Arg⁺Leu⁺ recombinants. Results are shown in the table below.

	A1	A2	A3	A4
L1	++	–	++	–
L2	–	++	–	+
L3	+	–	+	–
L4	–	++	–	+

++ = many Arg⁺Leu⁺ recombinants
+ = a few Arg⁺Leu⁺ recombinants
– = no Arg⁺Leu⁺ recombinants

Which of these strains is F⁺?

➤ **Solution**

Answer: Strains A4 and L3

The genotype of each strain (Hfr, F⁺, or F⁻) can be deduced by understanding the frequencies with which each of these strains transmits genes to the others. An Hfr × F⁻ cross yields large numbers of conjugants. ("Hfr" stands for *h*igh *f*requency of *r*ecombination). An F⁺ × F⁻ cross yields small numbers of conjugants. (The fertility factor of an F⁺ strain must first integrate into the donor chromosome before the resulting Hfr can transfer genes to an F⁻ recipient.) Hfr × Hfr, Hfr × F⁺, F⁺ × F⁺, and F⁻ × F⁻ crosses yield no recombinants. Thus, whenever large numbers of Arg⁺Leu⁺ recombinants are obtained, one parent is Hfr and one is F⁻. Whenever small numbers of Arg⁺Leu⁺ recombinants are obtained, one parent is F⁺ and one is F⁻. The puzzle is to decipher which one is which. Comparing results of different crosses aids the process. For example, L1 × A1 is an Hfr × F⁻ cross, and L3 × A1 is an F⁺ × F⁻ cross. Which strain is which? Strain A1 is involved in both crosses. An F⁻ recipient is involved in both crosses. Thus, strain A1 must be the F⁻. If A1 is F⁻, then L1 is Hfr, L3 is F⁺, and L2 and L4 are F⁻. Such logic identifies strains A2 and L1 as Hfr, A4 and L3 as F⁺, and A1, A3, L2, and L4 as F⁻.

12. Bacteriophage P1 (a generalized transducing phage) is grown on wild-type *E. coli* (*a⁺b⁺c⁺*). The resulting phage stock is used as a donor for three transduction experiments. For each experiment, the recipient strain is genotype *a⁻b⁻c⁻*.

Experiment #1: a^+ transductants are selected
Experiment #2: b^+ transductants are selected
Experiment #3: c^+ transductants are selected

Transductants arising in each of the experiments are tested for their inheritance of the other two donor markers. The following results are obtained:

EXPERIMENT #1		EXPERIMENT #2		EXPERIMENT #3	
$a^+b^-c^-$	87%	$b^+a^-c^-$	43%	$c^+a^-b^-$	21%
$a^+b^+c^-$	0%	$b^+a^+c^-$	0%	$c^+a^+b^-$	15%
$a^+b^-c^+$	10%	$b^+a^-c^+$	55%	$c^+a^-b^+$	60%
$a^+b^+c^+$	3%	$b^+a^+c^+$	2%	$c^+a^+b^+$	4%

(a) What is the relative order of a, b, and c?

(b) For each gene pair, the frequency of cotransduction can be independently determined in two of the above experiments. Indicate both of these frequencies.

$$a^+ - b^+ = \underline{\hspace{1cm}} \text{ and } \underline{\hspace{1cm}}$$
$$a^+ - c^+ = \underline{\hspace{1cm}} \text{ and } \underline{\hspace{1cm}}$$
$$b^+ - c^+ = \underline{\hspace{1cm}} \text{ and } \underline{\hspace{1cm}}$$

➤ **Solutions**

(a) *Answer:* Gene order = a-c-b (or the reverse, which is equivalent)

The fact that $a^+b^+c^+$ transductants are obtained demonstrates that all three genes are closely linked. Thus, a^+, b^+, and c^+ can all be inherited on a single transduced DNA fragment. What is the order of genes on this DNA? The relative order of a, b, and c can be deduced from the number of transductants of each genotype.

If, for example, the gene order is a-b-c, alleles a^+ and c^+ (but not b^+) can be inherited from the donor only by a quadruple crossover. Transduced DNA is linear, and the recipient chromosome into which it recombines is circular. Thus, an even number of reciprocal crossovers (2, 4, etc.) is required to maintain the circularity of the recipient chromosome. Four crossovers (one to the left of a, one between a and b, one between b and c, and one to the right of c) incorporate the donor's a^+ and c^+ alleles while retaining the recipient's b^- allele. Quadruple crossovers are less frequent than any double crossover. Thus, if the gene order is a-b-c, $a^+b^-c^+$ transductants are expected to be the least frequent class of recombinant when either a^+ or c^+ is selected. Is this true? Experiments #1 and #3 demonstrate that it is not. Thus, a-b-c is not the correct gene order. How about b-a-c? In this case, $b^+a^-c^+$ transductants are expected to be the least frequent class when either b^+ or c^+ is selected. Do the observations meet these expectations? Again, they do not (experiments #2 and #3). This leaves order a-c-b. In this case, $a^+c^-b^+$ transductants are expected to be the least frequent class when either a^+ or b^+ is selected. This is exactly what is observed (experiments #1 and #3). Thus, the gene order is *a-c-b*.

(b) *Answer:* Percent cotransduction:

$a^+ - b^+ = 3\%$ (Experiment #1) and 2% (Experiment #2)
$a^+ - c^+ = 13\%$ (Experiment #1) and 19% (Experiment #3)
$b^+ - c^+ = 57\%$ (Experiment #2) and 64% (Experiment #3)

Having determined the order of genes in part **(a)**, we can now calculate the percent cotransduction for each gene pair. When calculating the percent cotransduction between *a* and *b*, we can ignore gene *c*. From the cross $a^+b^+ \times a^-b^-$, the percent cotransduction can be determined in two ways:

1. the proportion of a^+b^+ transductants among total a^+ transductants $\times$ 100 (obtained from experiment #1), and
2. the proportion of a^+b^+ transductants among total b^+ transductants $\times$ 100 (obtained from experiment #2). In experiment #1, the percent cotransduction is $[^3/_{100}] \times 100 = 3\%$. In experiment #2, it is $[^2/_{100}] \times 100 = 2\%$.

When calculating the percent cotransduction between *a* and *c*, we can ignore gene *b*. From the cross $a^+c^+ \times a^-c^-$, the percent cotransduction can be determined in two ways:

1. the proportion of a^+c^+ transductants among total a^+ transductants $\times$ 100, (obtained from experiment #1), and
2. the proportion of a^+c^+ transductants among total c^+ transductants $\times$ 100, (obtained from experiment #3).

In experiment #1, the percent cotransduction is $[(10 + 3)/100] \times 100 = 13\%$. In experiment #3, it is $[(15 + 4)/100] \times 100 = 19\%$.

Similar calculations indicate that the percent cotransduction between *b* and *c* is 57% (experiment #2) and 64% (experiment #3).

13. You perform the following interrupted mating experiment:

Hfr($str^s ade^+ pro^+ glu^+ thi^+ arg^+$) $\times$ F$^-$($str^r ade^- pro^- glu^- thi^- arg^-$)

+ = prototrophy   *str* = streptomycin
− = auxotrophy   *ade* = adenine

r = resistant   *pro* = proline
s = sensitive   *glu* = glutamine
                    thi = thiamin
                    arg = arginine

The parentheses indicate that the order of these genes is unknown. These strains are mixed together and at the indicated times samples are withdrawn and shaken vigorously to disrupt any mating pairs. Samples are then plated on media that contain streptomycin plus four of the five nutritional requirements. Each of the selective media is lacking only one of the nutritional requirements

+ = colonies obtained
− = no colonies obtained

Time of Mating Interruption	SUPPLEMENT MISSING FROM THE SELECTIVE MEDIA				
	Proline	Adenine	Arginine	Glutamine	Thiamin
6 min.	−	−	−	+	−
11 min.	+	−	−	+	−
17 min.	+	−	−	+	+
23 min.	+	−	−	+	+
27 min.	+	+	−	+	+
32 min.	+	+	−	+	+
36 min.	+	+	+	+	+
40 min.	+	+	+	+	+

Draw a map of this region of the chromosome indicating the relative positions of all genes and the order in which they are transferred by the Hfr. For each gene, indicate its approximate location in minutes relative to the origin of Hfr transfer.

Answer

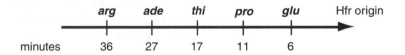

	arg	*ade*	*thi*	*pro*	*glu*	Hfr origin
minutes	36	27	17	11	6	

14. You cross a set of five phage mutants, *m1* through *m5*, in all possible combinations. You score the ability of each mutation to complement and to recombine with each other mutation. Note: one or more of these mutations may be a deletion.

Complementation Results
+ = complementation is observed
− = complementation is not observed

	m1	m2	m3	m4	m5
m1	−	+	−	+	−
m2		−	−	−	+
m3			−	−	−
m4				−	+
m5					−

Recombination Results
+ = recombination is observed
− = recombination is not observed

	m1	m2	m3	m4	m5
m1	−	+	+	+	+
m2		−	+	+	+
m3			−	−	−
m4				−	+
m5					−

(a) How many genes are defined by the six mutations?

(b) Which mutations affect which genes?

(c) Draw a genetic map for this phage. Be sure to indicate the position of any deletion endpoints relative to the mutations and to the gene boundaries.

Answer

(a) 2 genes are defined by *m1* through *m5*.

(b) *m1* and *m5* affect *gene1*; *m2* and *m4* affect *gene2*; *m3* is a deletion that affects both *gene1* and *gene2*.

(c)

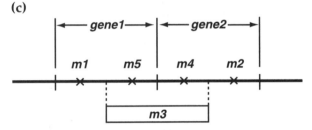

15. A stock of bacteriophage P22 is measured to contain 2 × 10⁸ pfu (plaque-forming units) per milliliter. One tenth milliliter of this stock is used to infect 5 × 10⁶ *E. coli* cells. Following one cycle of phage growth, the resulting lysate contains 1 × 10⁹ pfu. What is the multiplicity of infection in this experiment?

Answer 4

16. The map distances (in map units) between five alleles of the bacteriophage T4 *rII* gene are shown in the table on p. 155.

Map Distance Between *rII* Alleles (Map Units)

	r1	r2	r3	r4	r5
r1	<0.001				
r2	0.18	<0.001			
r3	0.08	0.12	<0.001		
r4	0.12	0.25	0.18	<0.001	
r5	0.20	0.03	0.14	0.30	<0.001

Construct a linkage map of these five alleles. Indicate the map distance between each allele and its nearest neighbor(s).

➤ *Solution*

Answer:

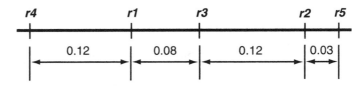

all distances in map units

Small map distances indicate mutations located close together, while larger map distances indicate mutations farther apart. The largest map distance in the table is 0.3 map units, between mutations *r4* and *r5*. Thus, *r4* and *r5* are the most distant mutations on the fine structure map. The smallest map distance in the table is 0.03 map units, between *r2* and *r5*. We have already established that *r5* is at one end of the linkage map. Thus, the relative order of these three mutations is *r4-r2-r5*. We can now look elsewhere in the table to determine the relative positions of *r1* and *r3*. *r1* and *r3* are 0.12 and 0.18 map units, respectively, from *r4* and 0.2 and 0.14 map units, respectively, from *r5*. Thus, the order of all five mutations is *r4-r1-r3-r2-r5*. The frequency of recombination between adjacent mutations can be read directly from the table.

17. Wild-type bacteriophage lambda (genotype λ *c⁺mi⁺*) forms large turbid plaques. Lambda *cI* and *mi* mutants form large clear and small turbid plaques, respectively. An *E. coli* host is coinfected with λ *cI mi⁺* and λ *c⁺ mi*. Progeny phage are

466 large clear, 454 small turbid, 43 large turbid, and 37 small clear plaques. What is the recombinant frequency between *cI* and *mi*?

➤ **Solution**

Answer: RF=0.08

Recombination frequency (RF) for a bacteriophage equals the number of recombinant progeny phage divided by the number of total progeny phage. From the cross λ *cI mi⁺* × λ *c⁺ mi*, recombinant progeny are genotype *c⁺mi⁺* (large turbid) and *cI mi* (small clear). Nonrecombinant progeny are genotype *cI mi⁺* (large clear) and *c⁺ mi* (small turbid). Thus,

$$RF = \frac{(43 + 37)}{(466 + 454 + 43 + 37)} = 0.08.$$

18. You cross a set of six phage mutants, *m1* through *m6*, in all possible combinations. You score the ability of each mutation to complement and to recombine with each other mutation. Note: one or more of these mutations may be a deletion.

Complementation Results
+ = complementation is observed
− = complementation is not observed

	m1	m2	m3	m4	m5	m6
m1	−	+	+	+	−	−
m2		−	−	+	−	+
m3			−	−	−	+
m4				−	−	+
m5					−	−
m6						−

Recombination Results
+ = recombination is observed
− = recombination is not observed

	m1	m2	m3	m4	m5	m6
m1	−	+	+	+	+	+
m2		−	+	+	+	+
m3			−	−	−	+
m4				−	−	+
m5					−	−
m6						−

(a) How many genes are defined by the six mutations?
(b) Which mutations affect which genes?
(c) Draw a genetic map for this phage. Be sure to indicate the position of any deletion endpoints relative to the mutations and to the gene boundaries.

Answer
(a) Three genes are defined by *m1* through *m6*.
(b) *m1* and *m6* affect *gene1*; *m4* affects *gene2*; *m2* affects *gene3*; *m3* is a deletion affecting both *gene2* and *gene3*; *m5* is a deletion affecting all three genes.

(c)

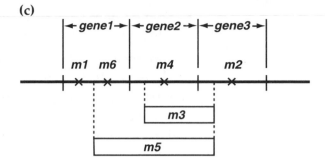

19. A stock of bacteriophage lambda is measured to contain 4×10^8 pfu (plaque-forming units) per milliliter. $1/10$ milliliter of this stock is used to infect 2×10^8 *E. coli* cells. Following one cycle of phage growth, the resulting 10 milliliter lysate contains 6×10^8 pfu per milliliter. What is the burst size in this experiment?

Answer 150

20. Wild-type bacteriophage λ (genotype h^+c^+) forms turbid plaques; λ *c*I and *c*II mutants form clear plaques. Wild-type λ infects only *E. coli* strain K12; λ *h* mutants infect either strains K12 or B. The *h* gene is 7 map units from either *c*I or *c*II. Genes *c*I and *c*II are located very close together. From the cross λ *c*I h^+ × λ *c*II h, 93% of the turbid-plaque progeny form plaques on either K12 or B; 7% form plaques only on K12. What is the order of the *c*I, *c*II, and *h* mutations on the phage λ linkage map?

Answer

h-*c*I-*c*II (or the reverse, which is equivalent)

21. A generalized transducing phage is grown on an *E. coli* strain that is $a^-b^+c^+$. This phage lysate is used to infect a recipient strain that is $a^+b^-c^-$.

"+" = prototrophy
"–" = auxotrophy
a^- = requires amino acid A
b^- = requires amino acid B
c^- = requires amino acid C

The relative order of the genes is *a*-*b*-*c* (*b* is in the middle). Phage-infected cells are plated on Minimal+A+B medium. One hundred colonies are obtained. Each of these 100 colonies is individually tested by replica plating for its ability to grow on the following media:

Minimal+A: 40 of the 100 colonies grow; 60 do not grow
Minimal+B: 80 of the 100 colonies grow; 20 do not grow

(a) What is the percent cotransduction between genes *a* and *c*?

(b) What is the percent cotransduction between genes *b* and *c*?

➤ **Solutions**

(a) *Answer:* 20%

(b) *Answer:* 40%

The first step of the solution is to establish the genotypes of the transductants. The recipient of the cross is genotype $a^+b^-c^-$. Phage-infected cells are plated on Minimal+A+B media. Thus, c^+ is selected, and all transductants are c^+. What about genes *a* and *b*? Forty of the c^+ transductants grow on Minimal+A medium. These 40 are b^+; the remaining 60 are b^-. Eighty of the c^+ transductants grow on Minimal+B medium. These 80 are a^+; the remaining 20 are a^-.

From the cross $a^-b^+c^+ \times a^+b^-c^-$, cotransduction of *a* and *c* occurs when c^+ transductants inherit a^-, and cotransduction of *b* and *c* occurs when c^+ transductants inherit b^+. Thus, the frequency of cotransduction between genes *a* and *c* is 20%, and the frequency between genes *b* and *c* is 40%.

---

22. You have collected four temperature sensitive mutations affecting your favorite bacteriophage (*ts1, ts2, ts3,* and *ts4*). You coinfect bacteria with pairs of mutant phage and measure the resulting burst sizes. You perform these infections and incubations at the non-permissive temperature.

Infecting Phages	Burst Size
wild-type	200
ts1	<0.001
ts2	<0.001
ts3	<0.001
ts4	<0.001
ts1 + ts2	200
ts1 + ts3	200
ts1 + ts4	200
ts2 + ts3	5
ts2 + ts4	<0.001
ts3 + ts4	200

(a) How many different genes are defined by these four mutations?

(b) Which mutations affect which genes?

(c) One of these mutations is likely to be a deletion. Which one is it?

(d) Draw a genetic map of these four mutations that is consistent with the data. Indicate, if possible, the boundaries of any genes. (Note: there is more than one solution with this data. Indicate any map that is consistent with the data.)

Answer

(a) Three

(b) *ts1* affects *gene1*; *ts4* affects *gene2*; *ts3* affects *gene3*; *ts2* is a deletion that affects both *gene2* and *gene3*.

(c) *ts2*

(d)

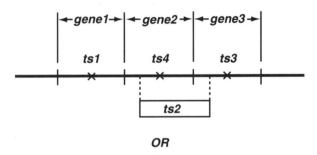

OR

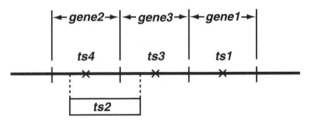

 23. The final two steps of synthesis of the amino acid histidine are shown in the diagram below.

$$\text{histidinol phosphate} \xrightarrow{\ hisB\ } \text{histidinol} \xrightarrow{\ hisD\ } \text{histidine}$$

hisD⁻ recipients are infected with generalized transducing phage grown on a *hisB⁻* donor. Phage-infected cells are placed on a minimal medium containing histidinol. One hundred *hisD*⁺ transductants are tested for their ability to grow on minimal medium: 12 grow, 88 do not. What is the percent cotransduction between *hisB* and *hisD*?

➤ *Solution*

Answer: 88%

Because *hisB⁻* mutants are *hisD⁺*, the growth requirements of *hisB⁻* mutants are satisfied either by histidine or by histidinol. *hisD⁻* mutants, on the other hand, cannot utilize histidinol as a nutritional source of histidine, because they cannot convert it to histidine.

The cross is *hisB⁻ hisD⁺* × *hisB⁺ hisD⁻*. Phage-infected cells are placed on medium containing histidinol. Thus, *hisD⁺* transductants are selected. If *hisD⁺* transductants *do not* inherit the donor's *hisB⁻* allele, their genotype is *hisD⁺ hisB⁺*. Such transductants are His⁺. If *hisD⁺* transductants *do* inherit the donor's *hisB⁻* allele, their genotype is *hisD⁺ hisB⁻*. Such transductants are His⁻, requiring either histidine or histidinol as a source of histidine. Twelve transductants are *hisD⁺hisB⁺*, and 88 are *hisD⁺ hisB⁻*. Thus the percent cotransduction between *hisD* and *hisB* is 88%.

24. You are mapping five *E. coli* genes using generalized transductional crosses. The genes are designated *A, B, C, D,* and *E*. You perform a series of crosses and measure the percent cotransduction between all possible pairs of genes. The results are shown in the table below.

Percent Cotransduction Between Gene Pairs

	A	B	C	D	E
A	–				
B	55	–			
C	14	32	–		
D	80	65	22	–	
E	45	25	4	34	–

Draw a genetic linkage map of genes *A* through *E*. Indicate the percent cotransduction between each gene and its nearest neighbor(s).

Answer

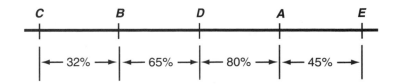

 25. As shown in the diagram below, *trpA*, *trpB*, and *trpC* are located close together, but their relative order is unknown. The frequency of cotransduction between each of these genes and *cysB* is approximately 50%.

trp⁻ cysB⁺ donors are used to transduce *trp⁻ cysB⁻* recipients, selecting Trp⁺ (Cys is unselected). The inheritance of Cys⁺ is scored among Trp⁺ recombinants. The results are shown in the following table.

Donor	Recipient	% Cys⁺
trpB⁻	*trpA⁻cysB⁻*	3%
trpA⁻	*trpC⁻cysB⁻*	50%
trpC⁻	*trpB⁻cysB⁻*	3%

What is the relative order of these four genes?

Answer *trpA-trpB-trpC-cysB*

topic 8
Genes and Proteins

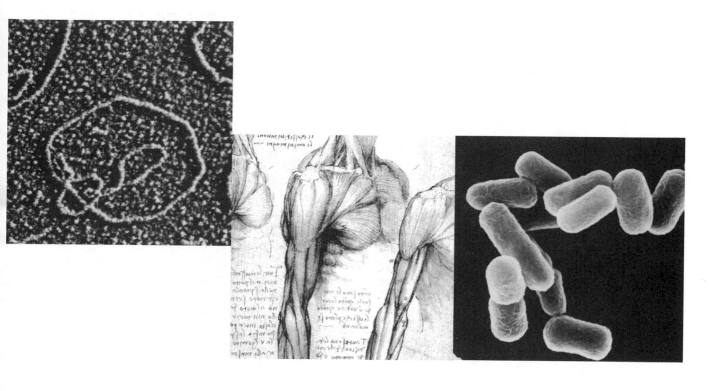

Summary

From the rediscovery of Mendel's Laws in 1900 until the 1940s, the question of how genes determine an organism's phenotype remained a mystery. Today we understand that virtually all phenotypic characteristics of an organism are governed by the activities of particular **proteins**. Even the simplest cells contain thousands of different proteins. Proteins are the true workhorses of cells, playing the most diverse roles of any class of molecules within living systems. Most proteins function as **enzymes** that catalyze biochemical reactions. Others have structural roles, act as hormones, are involved in transport processes, or participate in immunological defense. Altogether, the proteins that compose an organism dictate most aspects of its form and function. Genes ultimately determine phenotypes by providing the information necessary to direct the synthesis of each protein.

Proteins are very large macromolecules with molecular weights ranging from several thousand to several million. Despite the great variety of proteins present in living systems, they are all composed of the same twenty **amino acid** subunits. Amino acids are linked end-to-end by **peptide bonds** to form **polypeptide chains**. Most polypeptides consist of hundreds to thousands of amino acids. The exact sequence of amino acids in a polypeptide chain is a characteristic feature of a given protein and is what distinguishes one protein from another. The overall structure of proteins is described at four levels of increasing complexity. **Primary structure** refers to the linear sequence of amino acids in a polypeptide chain. **Secondary structure** is the local conformation assumed by the polypeptide chain as it folds into structures such as α helical coils or β pleated sheets, which are stabilized by hydrogen bonds between neighboring amino acids. **Tertiary structure** describes the bending and folding of the polypeptide chain to form spherical or globular shapes. **Quaternary structure** refers to the assembly of two or more polypeptide chains into a functional protein. A key point is that the unique function of each protein depends critically on its precise three-dimensional structure. In turn, this structure is determined by the linear sequence of amino acids in the polypeptide chain(s). Finally, the linear sequence of amino acids in polypeptides is encoded by the linear sequence of nucleotides within genes. In general, each gene encodes a single polypeptide chain.

The usual effect of a **mutation** is to cause a single amino acid to be replaced by another amino acid somewhere in the affected polypeptide chain. The correlation between the location of a mutation within a given gene and the position of the affected amino acid in the encoded polypeptide establishes the principle of **colinearity** between the sequence of nucleotides in a gene and the sequence of amino acids in a polypeptide.

Mutational changes can alter the overall configuration of a protein in such a way that it is no longer able to carry out its cellular function. For example, a mutation in a gene encoding an enzyme can result in the failure of a biochemical reaction because the enzyme's catalytic activity has been lost or impaired. As demonstrated by the example of **sickle cell anemia**, which is caused by a mutation in the gene encoding a component of hemoglobin, a complex cascade of cellular and developmental events is often interposed between the primary effect of a mutation at the molecular level and its ultimate phenotypic manifestation at the organismal level.

In simple microorganisms, such as Neurospora, many nutritional mutations have been obtained in which various metabolic steps are blocked owing to defects in the enzymes that would normally catalyze these steps. From the few simple ingredients in a **minimal medium**, wild-type Neurospora can synthesize all the chemical compounds required for growth. **Nutritional mutants**, in contrast, lack the ability to produce some essential nutrient, and can only grow if the medium is supplemented with the needed compound or one of the chemical intermediates in its biosynthesis. By determining which compounds are able to restore growth of a nutritional mutant, the point of blockage in a metabolic pathway can often be inferred. For example, compounds that are able to restore growth to a nutritional mutant are presumed to represent chemical intermediates in the biosynthesis of the required nutrient that occur after the point of the metabolic block. If a known chemical intermediate is unable to restore growth, it must represent a step that occurs prior to the point of blockage. A set of mutations that are all defective in the synthesis of the same nutrient can be analyzed in this way to dissect the overall metabolic pathway into an ordered sequence of discrete steps.

Many human genetic diseases have been traced to mutational blocks in various metabolic pathways. Even when two mutations affect closely related steps in the same metabolic pathway, their respective phenotypic consequences can be quite different. For example, alkaptonuria and phenylketonuria (PKU) are hereditary conditions caused by mutations at different steps in the metabolism of the amino acids phenylalanine and tyrosine. Although alkaptonuria is a relatively benign condition, PKU can result in severe mental retardation. Fortunately, simple dietary intervention beginning at birth can alleviate the most severe consequences of PKU. One of the primary goals of medical genetics is to pinpoint the underlying biochemical defect in human genetic diseases. As illustrated by the example of PKU, this knowledge is often crucial for early diagnosis of genetic diseases and for designing appropriate therapies.

The **complementation test** is an experimental means of determining whether two independent recessive mutations represent defects in the same gene. If the phenotype of an individual heterozygous for both mutations is normal, the mutations complement and are inferred to affect different genes. If the phenotype of the heterozygote is mutant, the mutations fail to complement and are inferred to affect the same gene. The biochemical basis of the complementation test is straightforward: Two mutations that affect *different* genes affect the production of *different* proteins. Individuals heterozygous for both mutations have one normal (non-mutant) allele of each gene that can direct the production of functional proteins. Thus, the resulting phenotype is normal. On the other hand, two mutations that affect the *same* gene affect the production of the *same* protein. In this case, both alleles present in a heterozygous individual are mutant and no functional protein encoded by that gene can be produced. Thus, the resulting phenotype is mutant.

Self-Testing Questions

1. Parallel and antiparallel β pleated sheets are common structural motifs of globular proteins. The β sheet is built from β strands, with several such strands from different regions of a polypeptide chain interacting to form a β sheet. The β strands are aligned adjacent to each other in either a parallel or antiparallel orientation, such that hydrogen bonds can form between carbonyl (C = O) groups of one β strand with secondary amines (NH) of an adjacent β strand. Such hydrogen bonds stabilize an extended region of the polypeptide chain in a planar structure. A β sheet is an example of:

   (a) the primary structure of a protein.
   (b) the secondary structure of a protein.
   (c) the tertiary structure of a protein.
   (d) the quaternary structure of a protein.

   Answer (b) the secondary structure of a protein

2. The molecular structures of valine (Val) and threonine (Thr) are shown in the diagram below.

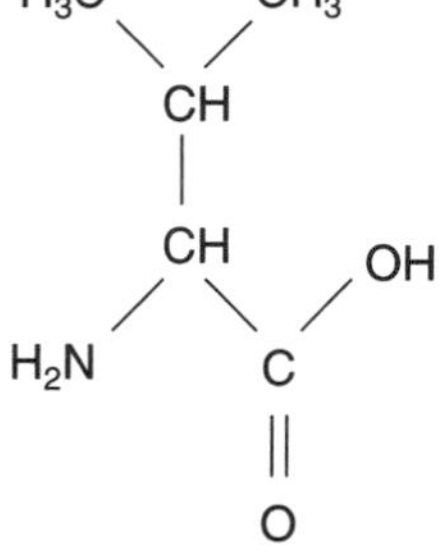

Val 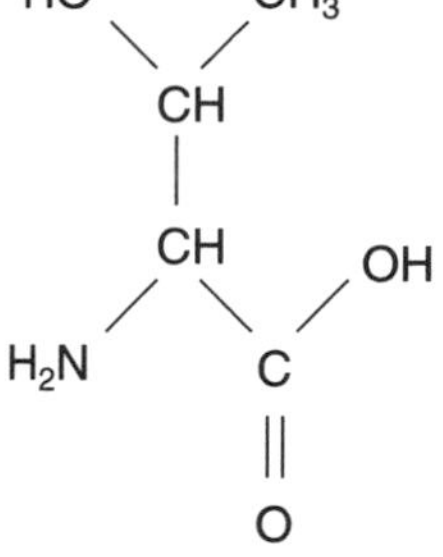   Thr

Draw the molecular structure of the dipeptide valyl-threonine, with valine being the amino-terminal amino acid and threonine the carboxyl-terminal amino acid.

Answer

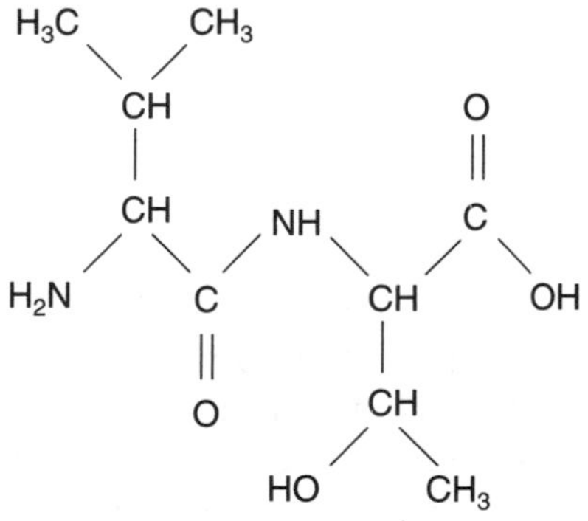

valyl-threonine

3. Alpha helices are common structural motifs of proteins. In α helices, amino acid n of a polypeptide chain forms noncovalent chemical bonds with amino acid $n+4$ of the chain. Such interactions stabilize an extended region of the polypeptide chain in a ribbon-like helical configuration. What type of noncovalent chemical bonds stabilize α helices?

Answer hydrogen bonds

4. The structure of an oligopeptide is shown in the diagram below.

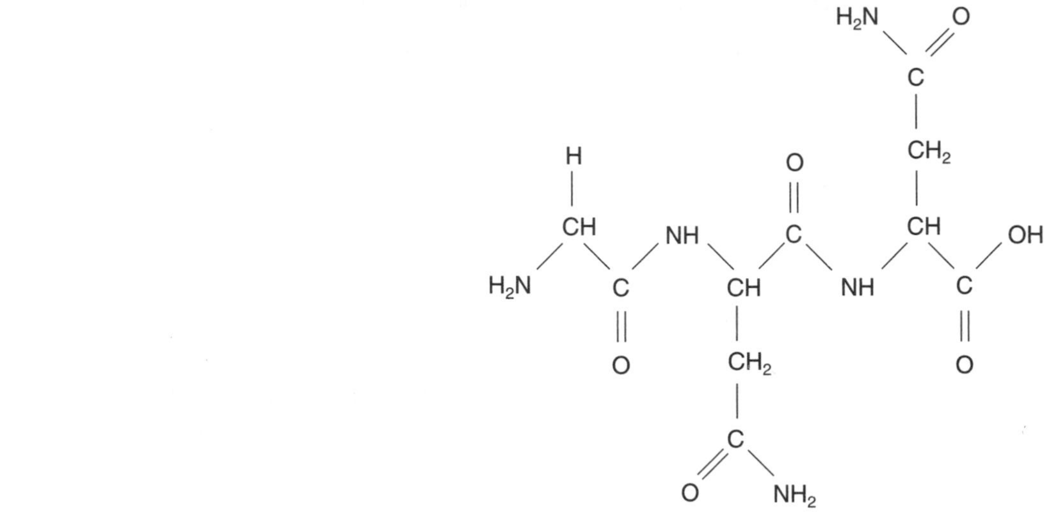

This oligopeptide contains how many amino acids?

Answer

Three. The oligopeptide shown has the sequence Gly-Asn-Asn.

5. The enzyme arginase hydrolyzes arginine to ornithine and urea. Protein extracts of beef liver are electrophoresed in a non-denaturing polyacrylamide gel and stained for arginase enzymatic activity. Such experiments indicate the sizes of active arginase enzyme in the sample. Individual cattle exhibit either one or four electrophoretically-separable bands of arginase activity. Assume that all differences in arginase electrophoretic mobility are due to allelic differences in the sequence of the arginase enzyme. Of how many identical subunits is arginase composed?

Answer 3

6. The molecular weight of circulating antibodies (immunoglobulin class IgG) is 145,200 g/mole. IgG molecules are tetramers, consisting of 2 identical immunoglobulin gamma

heavy chains complexed with 2 identical immunoglobulin light chains. IgG heavy chains are 440 amino acids each. (Although there is slight size variation among heavy chains, the variation is minor and can be ignored for the present.) How many amino acids are contained in IgG light chains? Assume that the average molecular weight of each amino acid is 110 g/mole.

➤ **Solution**

Answer: 220

The molecular weight of IgG (145,200 g/mole) represents the sum of the molecular weights of 2 molecules of heavy chain and 2 molecules of light chain. If we calculate the molecular weight contributions of the heavy chains, the remainder is that of the light chains. IgG heavy chains are 440 amino acids. Thus, the molecular weight of each heavy chain is $440 \times 110 = 48,400$ g/mole. Two such heavy chains are present in each IgG molecule. Thus, the molecular weight of each light chain is $[\frac{1}{2}][145,200 - (2 \times 48,400)] = 24,200$ g/mole. How many amino acids are contained in a protein of 24,200 g/mole? $24,200 \div 110 = 220$. Thus, each light chain is composed of 220 amino acids.

7. The enzyme glucose-6-phosphate dehydrogenase (G6PD) is encoded by a human X-linked gene, designated *Gd*. Numerous different *Gd* alleles exist worldwide that differ with regard to their intrinsic G6PD enzymatic activity (assayed in extracts of red blood cells). The genotypes and G6PD enzyme quantities of several individuals are shown in the table.

Genotype	G6PD Activity per Milligram of Protein (% Normal)
Gd^B / Gd^B	100%
$Gd^{Kerala} / Gd^{Kerala}$	50%
Gd^B / Gd^{Kerala}	75%
$Gd^{Kerala} / Gd^{Eyssen}$	25%

What quantity of G6PD is expected for individuals of genotype $Gd^{Eyssen} / Gd^{Eyssen}$?

➤ **Solution**

Answer: 0%

Many human genetic diseases are caused by a reduction or absence of a key metabolic enzyme. Such diseases are usually inherited as recessive traits. Homozygotes are affected, but heterozygotes are not. For such traits, carrier heterozygotes cannot usually be distinguished from homozygous normal individuals based on their phenotype (both are normal). In cases where a sensitive biochemical assay exists for the affected enzyme, heterozygotes can often be distinguished from normal individuals because they express a quantity of enzyme that is intermediate between normal and affected individuals. Although a trait may be *recessive* with regard to phenotype, it can be *codominant* with regard to the quantity of enzyme expressed in tissue samples.

G6PD is an example of such a trait. The quantity of G6PD enzyme in extracts of red blood cells reflects the genotype of the individual. Heterozygotes express a quantity of enzyme that is intermediate between that expressed by each of the respective homozygotes. For example, Gd^B/Gd^{Kerala} expresses 75% of normal, while Gd^B/Gd^B and Gd^{Kerala}/Gd^{Kerala} express 100% and 50%, respectively. The heterozygote, therefore, expresses approximately the average of the two homozygotes. Gd^{Kerala}/Gd^{Eyssen} heterozygotes express 25% of normal. Gd^{Kerala}/Gd^{Kerala} homozygotes express 50% of normal. How much enzyme is expected from Gd^{Eyssen}/Gd^{Eyssen} in order for 25% to be the average of 50% and Gd^{Eyssen}/Gd^{Eyssen}? The answer is 0%. Gd^{Eyssen}/Gd^{Eyssen} homozygotes are expected to produce little, if any, detectable enzyme.

Carriers of many metabolic diseases can be distinguished from homozygous normal individuals using the methods exemplified here, but the assignment of genotypes is sometimes uncertain. Due to both genetic and non-genetic effects, the quantity of enzyme expressed by individuals of differing genotype can be variable. The quantity of enzyme expressed by a given genotype usually falls within a range of values. The range exhibited by heterozygotes sometimes overlaps the range exhibited by one or both of the homozygotes. If the amount of enzyme expressed by an individual falls within the region of overlap, a genotype often cannot be established with certainty.

---

8. Galactosemia, an autosomal recessive disease, is caused by deficiency of the enzyme galactose-1-P uridyl transferase ("GALT"). Normal individuals (+/+), galactosemia homozygotes *(gt/gt)*, and galactosemia heterozygotes *(gt/+)* can be distinguished by the quantity of GALT enzyme in red blood cells:

Genotype	GALT Activity
+/+	28–40 units
gt/+	12–28 units
gt/gt	0–2 units

The following pedigree describes inheritance of galactosemia in two families. Numbers in the pedigree indicate the measured quantity of GALT enzyme activity.

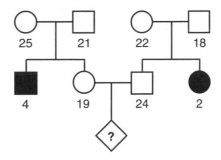

What is the probability that a child born to the indicated couple will be affected by galactosemia?

Answer 25%

9. The oxygen carrier protein hemoglobin is a tetramer composed of two α- and two β-globin polypeptide chains. The α- and β-globin genes are unlinked to each other. Inheritance of an autosomal recessive anemia is shown in the pedigree below. The α and β globin genotypes of all individuals are shown, where α1, α2, etc. indicate differing alleles of the α- and β-globin genes.

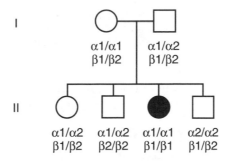

Is this pedigree consistent with the inherited anemia being caused by alleles of either α- or β-globin? Which one(s)?

➤ ***Solution***

Answer: This pedigree is consistent only with the inherited anemia being caused by the β1 allele.

The anemia in question is inherited as an autosomal recessive trait. Thus, affected individuals must be homozygous. The affected daughter (II.3) is homozygous for both alleles α1 and β1. Which, if either, of these alleles cause the anemia? Look through the pedigree for other individuals that are homozygous for either α1 or β1 (but not both). If those individuals are affected, the homozygous allele might be causative. If those individuals are not affected, the homozygous allele cannot be causative. The mother is homozygous for α1, yet she is unaffected. Thus, allele α1 cannot cause the anemia. Other than the affected daughter, no other β1 homozygotes are present in the pedigree. Thus, the pedigree is consistent with β1 being the cause of the inherited anemia.

10. *trp1*, *trp2*, *trp3*, and *trp5* mutants of yeast require the amino acid tryptophan because they are unable to synthesize it. All four of these genes are located on different chromosomes. A diploid of genotype *trp1/+ trp2/+ trp3/+ trp5/+* is induced to enter meiosis. What proportion of the haploid spores are expected to be tryptophan prototrophs?

    Answer $1/16$

11. The sequences of eight amino-terminal amino acids of four different β-hemoglobins are shown in the diagram below.

					POSITION			
	1	2	3	4	5	6	7	8
Hb #1:	Val	His	Leu	Thr	Pro	Val	Glu	Lys
Hb #2:	Val	Tyr	Leu	Thr	Pro	Glu	Glu	Lys
Hb #3:	Val	His	Leu	Thr	Pro	Lys	Glu	Lys
Hb #4:	Val	His	Leu	Thr	Pro	Glu	Glu	Lys

    One of these proteins is normal; the others are mutant alleles. Which of these β-hemoglobin sequences is normal?

    ➤ **Solution**

    Answer: Hb #4

    Each of the mutants is derived from normal by changing a single amino acid. Thus, the normal sequence must be related to all three mutant sequences by a single change. Two mutant sequences will often be related by two changes (one for each mutation). For example, Hb#1 and Hb#2 differ from each other at two positions. Neither of these can be the normal sequence, because two substitutions would be required to change one into the other. Which sequence is related to the other three by only a single change? Hb#4 differs from Hb#1 and Hb#3 only at position 6, and from Hb#2 only at position 2. Thus, Hb#4 is the normal sequence.

---

12. **a**-Factor is a small protein secreted by haploid yeasts of mating type "**a**" as part of their mating cycle. The amino acid sequences of seven overlapping peptide fragments of **a**-factor are shown in the diagram below. All peptide sequences are shown with the amino terminus to the left and the carboxyl terminus to the right.

    Peptide 1:   Phe-Trp
    Peptide 2:   Gly-Val
    Peptide 3:   Trp-Asp-Pro
    Peptide 4:   Tyr-Ile-Ile
    Peptide 5:   Pro-Ala-Cys
    Peptide 6:   Tyr-Ile-Ile-Lys-Gly
    Peptide 7:   Lys-Gly-Val-Phe-Trp-Asp

    What is the amino acid sequence of full length **a**-factor?

    ### Answer

    Tyr-Ile-Ile-Lys-Gly-Val-Phe-Trp-Asp-Pro-Ala-Cys

13. Hemophilia B is an X-linked recessive disease caused by mutations affecting blood coagulation Factor IX. Amino acids found at positions 180, 181, and 182 of Factor IX in three individuals affected with hemophilia B are shown in the table below.

	AMINO ACID POSITION		
	180	**181**	**182**
Allele #1	Arg	Val	Leu
Allele #2	Gln	Val	Val
Allele #3	Arg	Val	Phe

    What amino acids are found at these positions in normal individuals? Assume that hemophilia alleles are derived from normal alleles by single amino acid substitutions.

    ### ➤ Solution

    Answer: Arg-Val-Val

    The normal sequence is related to each of the mutant sequences by a single amino substitution. For each allele, therefore, two of the amino acids are normal and one is mutant. But which are normal and which is mutant? Position 182 is especially vexing. Three different amino acids are found at this position in the three alleles. Which, if any, of these is the normal sequence? We can deduce the normal amino acid at position 182 by inspecting position 180. Two alleles have Arg and a third has Gln at position 180. Because these alleles differ from normal at only one position, and because alleles 1, 2, and 3 differ at both sites 180 and 182, Arg must be the normal amino acid at position 180 (contained in alleles 1 and 3). If Arg is normal, then Gln is mutant, and position 182 of allele 2 (Val) must be normal. Thus, the normal amino acid sequence at these three positions is Arg-Val-Val.

14. Gonadotropin-releasing hormone (GRH) is a ten–amino acid peptide hormone that stimulates secretion of both luteinizing and follicle-stimulating hormones. The amino acid sequences of GRH from humans, alligators, and catfish are shown in the table below. (Single letter codes are used for amino acids.)

Source	Sequence
Human	Q-H-W-S-Y-G-L-R-P-G
Alligator	Q-H-W-S-Y-G-L-Q-P-G
Catfish	Q-H-W-S-H-G-L-N-P-G

Assume that these three GRHs were derived by mutation from a common evolutionary ancestor and that present-day GRHs contain only a single amino acid substitution relative to the ancestor. What is the sequence of GRH in the evolutionary ancestor?

Answer

Q-H-W-S-Y-G-L-N-P-G

15. Predatory snails of the genus *Conus* paralyze their victims by injecting them with α-conotoxins. Amino acid sequences of six overlapping peptide fragments of α-conotoxin from *C. geographus* and *C. striatus* are shown in the table below. All peptide sequences are shown with the amino terminus to the left and the carboxyl terminus to the right. Single-letter codes are used for amino acids.

	C. geographus	*C. striatus*
Peptide 1:	F-S-C	F-D-C
Peptide 2:	PACG	PACG
Peptide 3:	ECCH	YCCH
Peptide 4:	CGKH	CGKN
Peptide 5:	KHFSC	KNFDC
Peptide 6:	ECCHPA	YCCHPA

(a) What are the sequences of α-conotoxin from each of these species?

(b) Assuming that both of these sequences are derived from a common evolutionary ancestor, what is the minimum number of amino acid substitutions required to derive both *C. geographus* and *C. striatus* from the common ancestor?

Answer

(a) *C. geographus* = E-C-C-H-P-A-C-G-K-H-F-S-C
    C. striatus = Y-C-C-H-P-A-C-G-K-N-F-D-C

(b) three mutations for each species

16. The final step of histidine biosynthesis is conversion of histidinol to histidine by the enzyme histidinol dehydrogenase. Histidinol dehydrogenase in bacteria is encoded by the *hisD* gene. *HisD* mutants, therefore, are unable to utilize histidinol as a nutritional source of histidine. Mutations affecting any other histidine biosynthetic genes utilize either histidine or histidinol, because they are *hisD*$^+$. The growth properties of five bacterial mutants are shown in the table below.

	GROWTH AT 30°C			**GROWTH AT 37°C**		
	Min	Min+Hol	Min+His	Min	Min+Hol	Min+His
Wild-type	+	+	+	+	+	+
Mutant 1	−	−	+	+	+	+
Mutant 2	−	−	−	+	+	+
Mutant 3	+	+	+	−	−	+
Mutant 4	+	+	+	−	+	+
Mutant 5	−	+	+	+	+	+

Min = minimal medium; Hol = histidinol; His = histidine
+ = growth; − = no growth

Which of these mutants contains a heat-sensitive allele of *hisD*?

Answer Mutant 3

17. A simplified pathway of adenine biosynthesis in yeast is shown in the diagram on p. 176 (intermediates are abbreviated).

$$\text{FGAR} \xrightarrow{\textit{ADE6}} \text{FGAM} \xrightarrow{\textit{ADE7}} \text{AIR} \xrightarrow{\textit{ADE2}} \text{CAIR} \xrightarrow{\textit{ADE1}} \longrightarrow \text{Adenine}$$

The biosynthetic intermediate AIR is colored red. *ade2* mutants, which accumulate AIR, form red colonies. *ade1*, *ade6*, and *ade7* mutants, which do not accumulate AIR, form white colonies. All four genes segregate independently. A diploid of genotype *ade1/+ ade2/+ ade6/+ ade7/+* is induced to go through meiosis. What proportion of the haploid spores are expected to form white colonies?

➤ **Solution**

Answer: $\frac{7}{8}$

Red colonies are formed due to accumulation of the biosynthetic intermediate AIR. Only strains of genotype $ADE6^+$ $ADE7^+$ $ade2^-$ form red colonies. $ade6^-$ and $ade7^-$ single mutants, and $ade6^-$ $ade2^-$ or $ade7^-$ $ade2^-$ double mutants form white colonies. (In such mutants, AIR is not synthesized.) What proportion of haploid spores are genotype $ADE6^+$ $ADE7^+$ $ade2^-$? One half are $ADE6^+$; ½ are $ADE7^+$; ½ are $ade2^-$. Thus, (½)(½)(½) = ⅛ of the spores are expected to form red colonies, and $\frac{7}{8}$ are expected to form white colonies.

18. Two alternative pathways for biosynthesis of purple flower pigment in plants are shown in the diagram below.

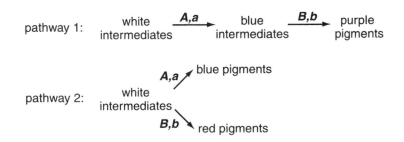

In pathway 1, white-colored intermediates are converted to blue intermediates by action of the *A,a* gene, and blue intermediates are subsequently converted to purple pigments by action of the *B,b* gene. In pathway 2, white-colored intermediates are converted to both blue pigments and red pigments by action of the *A,a* and *B,b* genes, respectively.

Purple color in pathway 2 results from the mixing together of red and blue pigments.

Alleles *a* and *b* are recessive mutations that eliminate enzyme function; alleles *A* and *B* are their normal counterparts. The *A*,*a* and *B*,*b* genes segregate independently. A plant of genotype *Aa Bb* is self-fertilized. For each pathway, what flower colors are expected among the offspring, and in what proportions are they expected?

Answer

Pathway 1: 9:4:3 (purple:white:blue)
Pathway 2: 9:3:3:1 (purple:red:blue:white)

19. The yeast "*ade5,7*" gene encodes a bifunctional enzyme that catalyzes steps 2 and 5 of adenine biosynthesis. Mutations of *ade5,7* can inactivate either enzymatic activity alone, or they can inactivate both activities together. Six different haploid adenine auxotrophs (ade^A–ade^F) are crossed in pairs. Phenotypes of the resulting diploids are shown in the table below.

	ade^A	ade^B	ade^C	ade^D	ade^E	ade^F
ade^A	Ade⁻					
ade^B	Ade⁺	Ade⁻				
ade^C	Ade⁺	Ade⁺	Ade⁻			
ade^D	Ade⁺	Ade⁻	Ade⁻	Ade⁻		
ade^E	Ade⁻	Ade⁺	Ade⁺	Ade⁺	Ade⁻	
ade^F	Ade⁺	Ade⁺	Ade⁻	Ade⁻	Ade⁺	Ade⁻

Which of these mutations is an allele of *ade5,7* that eliminates both enzymatic activities?

➤ Solution

Answer: ade^D

Complementation tests are designed to test whether two mutations affect the same gene or different genes. When two mutations fail to complement, we infer that they affect the same gene. When they complement, we infer that they affect different genes. In the case of the *ade5,7* "gene," however, we must modify the concept of complementation to be more precise. When two mutations fail to complement, they affect the same *biochemical activity*, and when they complement, they affect different *biochemical activities*. *Ade5,7* encodes two biochemical activities, which we will call Activity A and Activity B. If a mutation that destroys Activity A but retains Activity B (A⁻B⁺) is made heterozygous to a mutation that destroys Activity B but retains Activity A (A⁺B⁻), the diploid is expected to be phenotypically A⁺B⁺ (both activities are present). Thus, in special cases like *ade5,7*, a single gene can have two complementation groups.

A mutation that destroys both Activities A and B is expected to belong to both complementation groups. Thus, to identify an *ade5,7* allele that eliminates both enzymatic activities, we must sort the alleles into complementation groups and identify an allele that belongs to two different groups. In the table above, *ade/ade* diploids that are Ade⁻ fail to complement, and those that are Ade⁺ complement. These data indicate that the six *ade* alleles define three complementation groups. Alleles *ade^A* and *ade^E* define one complementation group, alleles *ade^B* and *ade^D* another, and alleles *ade^C*, *ade^D*, and *ade^F* a third. Note that *ade^D* fails to complement alleles of two different groups. Thus, *ade^D* is the *ade5,7* allele that eliminates both enzymatic activities.

20. Eye color in Drosophila is determined by two separate pathways that produce either brown or red eye pigment. *scarlet* (*st*) and *cinnabar* (*cn*) mutants are unable to synthesize brown pigment. As shown in the table below, *st* and *cn* single mutants and *st cn* double mutants accumulate either 3-hydroxykynurenine (3-HKU) or kynurenine (KU), both of which are intermediates in synthesis of brown pigment.

Genotype	Intermediate Accumulated
st	3-HKU
cn	KU
st cn	KU

What is the order of biosynthetic steps in the synthesis of brown pigment?

Answer

KU → 3-HKU → brown pigment

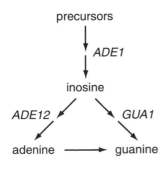

21. A simplified pathway for biosynthesis of purines in yeast is shown in the diagram to the left.

Cells must have an adequate supply of both adenine and guanine for growth. Note that adenine can be converted to guanine, but guanine cannot be converted to adenine. Fill in "+" or "−" in the table below to indicate whether the following genotypes will grow on the indicated growth media.

Min = minimal medium
Ino = inosine
Ade = adenine
gua = guanine

Genotype	Min	Min+Ino	Min+Ade	Min+Gua	Min+Ino+Gua
ade1	___	___	___	___	___
gua1	___	___	___	___	___
ade12	___	___	___	___	___
ade1 gua1	___	___	___	___	___
ade1 ade12	___	___	___	___	___

Answer

Genotype	Min	Min+Ino	Min+Ade	Min+Gua	Min+Ino+Gua
ade1	−	+	+	−	+
gua1	−	+	+	+	+
ade12	−	−	+	−	−
ade1 gua1	−	+	+	−	+
ade1 ade12	−	−	+	−	−

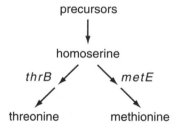

precursors

homoserine

thrB   *metE*

threonine   methionine

22. A simplified pathway for biosynthesis of the essential amino acids threonine and methionine in *E. coli* is shown in the diagram to the left.

You place 10^8 cells of genotype *thrB⁻ metE⁻* (double mutant) on minimal medium containing homoserine and methionine. Ten colonies (revertants) grow. The growth properties of these ten revertants are tested by replica plating. Fill in "+" or "−" in the table below to indicate whether the revertants will grow on the indicated growth media.

Min = minimal medium
Hom = homoserine
Thr = threonine
Met = methionine

Min	Min+Hom	Min+Thr	Min+Met
___	___	___	___

Answer

Min	Min+Hom	Min+Thr	Min+Met
−	−	−	+

23. The biosynthetic pathway of the amino acid arginine and the yeast genes that encode the biosynthetic enzymes are shown in the diagram below.

$$\text{ornithine} \xrightarrow{\textit{arg3}} \text{citrulline} \xrightarrow{\textit{arg1}} \begin{array}{c}\text{arginino-} \\ \text{succinate}\end{array} \xrightarrow{\textit{arg4}} \text{arginine}$$

All three genes segregate independently. A diploid of genotype $arg1^-/+ \ arg3^-/+ \ arg4^-/+$ is induced to go through meiosis. What proportion of the haploid spores are expected to grow on minimal + citrulline medium?

➤ *Solution*

Answer: ¼

Any strain that is *ARG1⁺ ARG4⁺* is able to grow on minimal medium supplemented with citrulline. When an *arg1⁻/+ arg4⁻/+* diploid goes through meiosis, what proportion of haploid spores are *ARG1⁺ ARG4⁺*? One-half are *ARG1⁺*, and one-half are *ARG4⁺*. Thus, (½)(½) = ¼ of the spores are both *ARG1⁺* and *ARG4⁺*. Such haploid strains grow on minimal medium containing citrulline.

24. Wild-type yeast cannot utilize α-aminoadipate (α-AA) as a sole source of nutritional nitrogen, but *lys2* mutants (which require lysine) can utilize α-AA as a sole nitrogen source. Lysine auxotrophs other than *lys2* mutants are unable to utilize α-AA as a nitrogen source. 10^7 cells of a lysine auxotroph that cannot utilize α-AA as a sole nitrogen source are plated on media containing lysine and α-AA as the sole nitrogen source. Ten colonies grow. One of these colonies is purified and crossed with wild type. The resulting diploid is purified and induced to go through meiosis. What proportion of the resulting haploids are expected to grow on minimal medium containing a nitrogen source other than α-AA? Assume that all lysine biosynthetic genes segregate independently.

Answer ¼

25. The pathway of biosynthesis of the amino acid methionine in yeast, and the growth properties of yeast methionine auxotrophs are shown below.

homoserine ──────▶ acetylhomoserine ──────▶ homocysteine ──────▶ methionine
(HS)   (AHS)   (HC)   (MET)

	MINIMAL MEDIUM PLUS:			
Genotype	AHS	HC	HS	MET
met2	+	+	−	+
met3	+	+	+	+
met6	−	−	−	+
met25	−	+	−	+

+ = growth; − = no growth

Which genes encode which enzymes?

➤ *Solution*

Answer:

                MET3   MET2   MET25   MET6
precursors ──────▶ HS ──────▶ AHS ──────▶ HC ──────▶ MET

Mutants with blocks in biochemical pathways are expected to assimilate and utilize pathway intermediates that occur *after* the block, but not those that occur *before* the block. In the methionine pathway, for example, mutants blocked in the final biosynthetic step (conversion of homocysteine to methionine) are expected to be unable to utilize any of the biosynthetic intermediates as a source of methionine. Even if such intermediates are present in the growth medium, they cannot be converted into methionine. *met6* mutants in the table above have these growth properties. Thus, *MET6* encodes the enzyme that converts homocysteine to methionine. Similarly, mutants blocked in the conversion of acetylhomoserine (AHS) to homocysteine (HC) are expected to utilize HC as a source of methionine, but not AHS or earlier intermediates. *met25* mutants have these growth properties. *MET25*, therefore, encodes the enzyme that converts AHS to HC. Using similar logic, we conclude additionally that *MET2* is required to convert homoserine to AHS, and that *MET3* is required for synthesis of homoserine from unspecified precursors.

26. A hypothetical pathway for biosynthesis of amino acid "E" is shown in the following diagram.

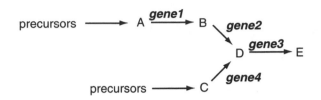

Enter "+" or "−" in the following table to indicate whether the following mutants are expected to grow on the indicated media. Assume that intermediate compounds A, B, C, and D can all be transported into the cell if supplied in the growth medium. Min = minimal medium.

Genotype	Min + A	Min + B	Min + C	Min + D	Min + E
gene1⁻	——	——	——	——	——
gene2⁻	——	——	——	——	——
gene3⁻	——	——	——	——	——
gene4⁻	——	——	——	——	——
gene1⁻gene2⁻	——	——	——	——	——
gene1⁻gene4⁻	——	——	——	——	——

Answer

Genotype	Min + A	Min + B	Min + C	Min + D	Min + E
gene1⁻	+	+	+	+	+
gene2⁻	+	+	+	+	+
gene3⁻	−	−	−	−	+
gene4⁻	+	+	+	+	+
gene1⁻gene2⁻	+	+	+	+	+
gene1⁻gene4⁻	−	+	−	+	+

 27. Wild-type Drosophila have bright red eyes. *white, garnet,* and *apricot* are X-linked recessive mutations that alter eye color. Offspring resulting from crosses of *white* females × *garnet* males and *white* females × *apricot* males are shown in the table below.

PARENTS		OFFSPRING	
Female	**Male**	**Female**	**Male**
white	*garnet*	red	white
white	*apricot*	apricot	white

Which, if any, of these eye color mutations affect the same gene?

➤ *Solution*

Answer: white and *apricot* are alleles of the same gene. *garnet* is an allele of a different gene.

Drosophila have an XX/XY system of sex determination. Thus, tests of complementation can only be done in females. The question states that *white*, *garnet*, and *apricot* are recessive. Thus, females of genotype *white/+*, *garnet/+*, or *apricot/+* have red eyes. What about *white/garnet* and *white/apricot* females? *white/garnet* females have red eyes. Such heterozygotes must express both *white*[+] and *garnet*[+] activity. Thus, *white* and *garnet* complement each other and likely affect different genes. *white/apricot* females, on the other hand, have apricot eyes. Thus, *white* and *apricot* fail to complement and likely affect the same gene.

---

28. You have isolated six mutants of haploid yeast that require the amino acid tryptophan (trp^A through trp^F). You cross these mutants in pairs and determine whether the resulting diploids are Trp$^+$ or Trp$^-$. Results are shown in the table below.

	trp^A	trp^B	trp^C	trp^D	trp^E	trp^F
trp^A	Trp$^-$					
trp^B	Trp$^+$	Trp$^-$				
trp^C	Trp$^+$	Trp$^-$	Trp$^-$			
trp^D	Trp$^-$	Trp$^+$	Trp$^+$	Trp$^-$		
trp^E	Trp$^-$	Trp$^+$	Trp$^+$	Trp$^-$	Trp$^-$	
trp^F	Trp$^-$	Trp$^+$	Trp$^+$	Trp$^-$	Trp$^-$	Trp$^-$

How many genes are defined by the six tryptophan auxotrophs?

Answer

Two. Trp^A, trp^D, trp^E, and trp^F affect one gene; trp^B and trp^C affect a second gene.

---

29. Five mutants of your favorite flowering plant have white, rather than the normal red, flowers. The mutations causing white flowers are designated *w1* through *w5*. When self-fertilized, all five mutants are true-breeding, yielding only white-flowered offspring. Mutants *w1* through *w5* are crossed in pairs, and the flower color of the F1 hybrids are scored. Results are shown in the table below.

Flower Color of F1 Hybrids

	w1	w2	w3	w4	w5
w1	White				
w2	Red	White			
w3	White	Red	White		
w4	Red	Red	Red	White	
w5	Red	Red	Red	White	White

Each of the F1 hybrids is allowed to self-fertilize. All white-flowered F1's yield only white-flowered F2. Red-flowered F1's, however, yield both red- and white-flowered F2. Two different classes of red-flowered F1 are distinguished: Hybrids *w1/w4*, *w1/w5*, *w3/w4*, and *w3/w5* yield a 1:1 ratio of red-flowered:white-flowered offspring. Hybrids *w1/w2*, *w2/w3*, *w2/w4*, and *w2/w5* yield a 9:7 ratio of red-flowered:white-flowered offspring.

(a) How many genes are defined by *w1* through *w5*?
(b) Which alleles affect which genes?
(c) Are any of the genes linked? If so, which ones?

➤ *Solutions*

(a) *Answer:* Alleles *w1* through *w5* define three genes.

Each of the F1 hybrids represents a complementation test between two white-flowered alleles. If a *w/w* hybrid has white flowers, the two mutations involved fail to complement and, therefore, affect the same gene. If a *w/w* hybrid has red flowers, the two mutations involved complement and, therefore, affect different genes. Alleles *w1* and *w3* fail to complement, as do alleles *w4* and *w5*. All other combinations complement. Thus, three genes are defined by alleles *w1* through *w5*.

(b) *Answer:* Alleles *w1* and *w3* affect *gene1*; alleles *w4* and *w5* affect *gene2*; allele *w2* affects *gene3*.

Alleles that fail to complement affect the same gene. Alleles *w1* and *w3* affect a single gene, which we will call *gene1*. Alleles *w4* and *w5* affect a different gene, which we will call *gene2*. Allele *w2* affects a third gene, which we will call *gene3*.

(c) *Answer: gene1* and *gene2* are linked; *gene3* is unlinked to both *gene1* and *gene2*.

Red-flowered F1 are heterozygous for two different *white* mutations that affect two different genes. Such genes might be linked, or they might be unlinked. Consider two genes, which we will call *A,a* and *B,b*. (Alleles *a* and *b* are the recessive white-flowered alleles; *A* and *B* are the dominant red-flowered alleles.) If genes *A,a* and *B,b* are linked, an F1 hybrid of genotype *Ab/aB* yields offspring that are:

$\frac{1}{4}$ *Ab/Ab* homozygotes
$\frac{2}{4}$ *Ab/aB* heterozygotes
$\frac{1}{4}$ *aB/aB* homozygotes

Both classes of homozygotes (*Ab/Ab* and *aB/aB*) have white flowers; heterozygotes (*Ab/aB*) have red flowers. Thus, genes that are tightly linked yield a 1:1 ratio of red-flowered:white-flowered offspring. What about unlinked genes?

If *A,a* and *B,b* are unlinked, an F1 hybrid of genotype *Aa Bb* yields offspring that are:

$\frac{9}{16}$ *A- B-*
$\frac{3}{16}$ *aa B-*
$\frac{3}{16}$ *A- bb*
$\frac{1}{16}$ *aa bb*

Only plants of genotype *A- B-* have red flowers; all others have white flowers. Thus, unlinked genes yield a 9:7 ratio of red-flowered:white-flowered offspring.

We deduce from this analysis that linked genes yield a 1:1 ratio of offspring, while unlinked genes yield a 9:7 ratio. Allele pairs *w1/w4, w1/w5, w3/w4,* and *w3/w5* all yield a 1:1 ratio of offspring. Thus, they are linked. Alleles *w1* and *w3* affect *gene1*; alleles *w4* and *w5* affect *gene2*. All F1 hybrids involving allele *w2*, which defines *gene3*, exhibit a 9:7 ratio of F2. Thus, *gene3* is unlinked to either *gene1* or *gene2*.

DNA Replication, Repair & Mutation

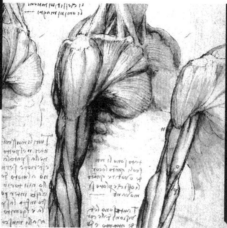

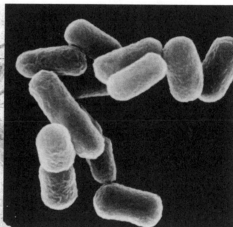

Summary

The essential properties of genes as determined from classical breeding studies are:

1. they store the information necessary to direct the production of phenotypes.
2. they are capable of precise self-replication.
3. they undergo sporadic mutational changes resulting in altered forms that are replicated as faithfully as the original.

Although the molecular basis of these properties remained a mystery until the 1950s, the discovery that **DNA** is the genetic material and the elucidation of the structure of this molecule ultimately provided the key insights into how genes work.

DNA is a right-handed **double helix** composed of two **polynucleotide chains** that run in opposite or **antiparallel** orientation to each other. The elementary subunit of DNA is a **nucleotide** consisting of a sugar **(deoxyribose)**, a nitrogen-containing base **(adenine, guanine, cytosine, or thymine)**, and a phosphate group attached to the 5' carbon of the sugar. In polynucleotide chains, **phosphodiester bonds** link the phosphate ($-PO_4$) group at the 5' end of one nucleotide with the hydroxyl (-OH) group at the 3' end of the preceding nucleotide. Thus, each polynucleotide chain has distinct 5' and 3' ends with a phosphate group at the 5' end and a hydroxyl group at the 3' end. The sugar-phosphate backbone lies along the outside of the DNA molecule with the bases pointing inwards. The most important feature of DNA is the specific pairing of bases via the formation of hydrogen bonds between **complementary** bases. Adenine pairs with thymine and guanine pairs with cytosine. This complementary base pairing underlies all the biological functions of DNA.

Each gene has its own characteristic linear nucleotide sequence that contains the information necessary to direct the production of the polypeptide it encodes. For this information to be transmitted faithfully from cell to cell and from generation to generation, each DNA molecule must replicate itself accurately. The general mode of DNA replication is **semi-conservative**; each of the two parental strands serves as a template for the assembly of a new daughter strand. Along each **template**, appropriate nucleotides are aligned by complementary base pairing and then linked together by formation of phosphodiester bonds. After replication, two duplicate

DNA molecules are produced, each consisting of one parental strand and one newly synthesized strand.

Replication is initiated at specific sites called **origins of DNA replication**. Typically, phage or bacterial chromosomes have a single origin of replication, whereas eukaryotic chromosomes, which consist of much longer DNA molecules, utilize multiple origins of replication. Replication in both cases is usually **bidirectional**; two **replication forks** advance in opposite directions from each origin of replication. Replication requires the activity of **DNA polymerase**, an enzyme that directs the growth of a polynucleotide chain by catalyzing the formation of phosphodiester bonds between adjacent nucleotides. All known DNA polymerases are able to catalyze the growth of a polynucleotide chain only in the 5′ to 3′ direction. In other words, a polynucleotide chain can only grow by the addition of incoming nucleotides at its 3′ end via the formation of a phosphodiester bond between the 5′-PO$_4$ of an incoming nucleotide and the free 3′-OH of the growing strand.

Because of the specific polarity of DNA polymerase and because of the antiparallel orientation of the two DNA strands, only one of the two daughter strands being synthesized at a replication fork is extended in the same direction as the replication fork is advancing. This strand, referred to as the **leading strand**, is oriented such that its 3′ end points toward the replication fork. This strand is synthesized continuously by the addition of nucleotides at its 3′ end as the replication fork advances. The other newly made strand, referred to as the **lagging strand**, is oriented with its 3′ end pointing away from the replication fork. This strand is synthesized discontinuously by the production of a series of short fragments that are eventually linked together into a complete strand. Each of these fragments, called **Okazaki fragments**, is initiated at the replication fork and then extended in the 5′ to 3′ direction away from the replication fork. Initiation of DNA synthesis requires the activity of an enzyme called **primase** that synthesizes a short fragment of RNA that serves as a primer for DNA synthesis. Once the RNA primer is made, DNA polymerase (DNA polymerase III in *E. coli*) extends the chain by adding deoxyribonucleotides at its 3′ end. DNA synthesis continues until a newly made fragment abuts the RNA primer of the previously made Okazaki fragment. Another DNA polymerase (DNA polymerase I in *E. coli*) then removes the RNA sequence in front of the growing fragment and further extends this fragment until its 3′ end immediately adjoins the 5′ end of the previously made Okazaki fragment. Finally, these fragments are covalently linked together by **DNA ligase**, which catalyzes the formation of a phosphodiester bond between the adjoining chains. Initiation of leading strand synthesis also requires primase but, in contrast with the events just described for the lagging strand, the leading strand requires only one or a few such initiation events.

Spontaneous **mutations** may result from occasional mistakes in base pairing during replication, resulting in misincorporation of bases. However, under normal circumstances, the observed frequency of such errors in newly replicated DNA molecules is extremely low. The basis of this high fidelity is that when pairing errors occur, they are usually corrected almost immediately. For example, DNA polymerase III in *E. coli* has a **proofreading** or editing activity that enables it to monitor whether the nucleotides it has just added to a growing chain are correct. If a mistake is detected, the enzyme removes the erroneous nucleotide and tries again. Only those rare mistakes that escape detection will give rise to mutations.

Mutations may also arise in DNA following physical or chemical damage to one of the bases caused by exposure to agents such as various types of radiation or certain chemicals. For example, ethylmethane sulfonate causes chemical modifications of certain bases that are likely to mispair during the next round of replication. The accumulation of spontaneous or induced errors in DNA would become intolerably high if the integrity of DNA were not maintained by a variety of DNA **repair systems.** These systems continually monitor DNA molecules for lesions associated with mismatched or damaged bases. One general mechanism for repairing such lesions when they are encountered is the enzymatic removal of a small segment of the DNA strand containing the lesion followed by resynthesis of that segment using the remaining intact strand as a template. The importance of these repair mechanisms is revealed by the severe phenotypic consequences in organisms from bacteria to humans when one or more of these repair systems is lacking.

Another source of spontaneous mutations in most organisms is **transposable elements.** Transposable elements are small segments of DNA that occupy no fixed location in the genome but instead jump or transpose from one location to another, more or less at random. If such an element happens to land within or near a particular gene, the function of the gene may be disrupted, leading to a mutant phenotype. One characteristic of such mutations is that they are often unstable. That is, the mutations revert back to normal at relatively high frequencies. These reversion events occur when the element transposes to a new location, restoring the function of the previously disrupted gene.

Knowledge of the structure of DNA and the activities of enzymes involved in its replication and repair provide the basis for molecular models of **recombination.** Most such models propose that recombination involves an exchange of corresponding single-stranded segments between two homologous DNA molecules. This exchange creates regions of **heteroduplex DNA** in which a strand derived from one parental molecule is base paired with a strand from the other parental molecule. Following strand exchange, a cross bridge links the two heteroduplex DNA molecules into a joint structure. The process of **branch migration** in this joint molecule moves the position of the cross bridge between the molecules. As the bridge moves, the heteroduplex sections of DNA vary in extent and location. Eventually, the linked molecules are resolved by enzyme-mediated cutting and religation into separate DNA double helices containing segments of DNA contributed by each of the two parental molecules. These molecular models account for most of the known experimental details of recombination and are supported by direct electron microscopic observations of joint heteroduplex structures in recombining DNA molecules.

Self-Testing Questions

1. Indicate whether the following statements are true or false.

    (a) Double-stranded DNA contains 10.4 base pairs per helical turn.

    (b) The sugar of DNA's sugar-phosphate backbone is a pentose sugar.

    (c) DNA polymerase I is required for replication of the *E. coli* chromosome.

    (d) DNA replication is initiated on both the leading and lagging strands by a short RNA primer.

    (e) The enzyme DNA helicase relieves the superhelical strain that builds up in double-stranded DNA as it is unwound during DNA replication.

    (f) Okazaki fragments are joined together by the enzyme DNA ligase.

_____ **(g)** A single polynucleotide chain in solution can have both single- and double-stranded regions within the same molecule.

_____ **(h)** The 5′→3′ polymerase activity of DNA polymerase is required during branch migration of a Holliday structure.

_____ **(i)** According to the double-strand break repair model of homologous recombination, an intermediate is formed during recombination that contains two separate Holliday junctions.

_____ **(j)** A mutation that changes an A:T base pair to a G:C base pair is a transversion mutation.

_____ **(k)** Pyrimidine dimers in DNA, caused by exposure to ultraviolet light, are directly repaired by the enzyme DNA photolyase.

Answer

_F_ **(a)** Double-stranded DNA contains 10.4 base pairs per helical turn.

_T_ **(b)** The sugar of DNA's sugar-phosphate backbone is a pentose sugar.

_F_ **(c)** DNA polymerase I is required for replication of the *E. coli* chromosome.

_T_ **(d)** DNA replication is initiated on both the leading and lagging strands by a short RNA primer.

_F_ **(e)** The enzyme DNA helicase relieves the superhelical strain that builds up in double-stranded DNA as it is unwound during DNA replication.

_T_ **(f)** Okazaki fragments are joined together by the enzyme DNA ligase.

_T_ **(g)** A single polynucleotide chain in solution can have both single- and double-stranded regions within the same molecule.

_F_ **(h)** The 5′→3′ polymerase activity of DNA polymerase is required during branch migration of a Holliday structure.

_T_ **(i)** According to the double-strand break repair model of homologous recombination, an intermediate is formed during recombination that contains two separate Holliday junctions.

_F_ **(j)** A mutation that changes an A:T base pair to a G:C base pair is a transversion mutation.

_T_ **(k)** Pyrimidine dimers in DNA, caused by exposure to ultraviolet light, are directly repaired by the enzyme DNA photolyase.

2. In herring sperm DNA, 27.6% of the bases are thymine. What proportion of herring sperm bases are guanine?

Answer 22.4%

3. Is deoxyguanosine 5′-monophosphate (dGMP) a nucleotide or a nucleoside?

Answer dGMP is a nucleotide.

4. Bacteriophage genomes (those contained within the virus particle itself) are either single- or double-stranded. The base composition of eight bacteriophage DNAs are shown in the table below.

	BASE COMPOSITION			
Phage	**A**	**G**	**T**	**C**
Pf3	19.8%	24.1%	34.8%	21.3%
P4	25.2%	24.8%	25.2%	24.8%
C2	31.8%	18.2%	31.8%	18.2%
IKe	25.1%	21.6%	34.3%	18.9%
PZA	30.2%	19.8%	30.2%	19.8%
S13	23.9%	23.3%	31.7%	21.1%
fd	24.6%	20.7%	34.5%	20.2%
Cp-1	30.6%	19.4%	30.6%	19.4%

Which of these bacteriophage contain single-stranded DNA within the virus?

➤ *Solution*

Answer: Bacteriophages Pf3, IKe, S13, and fd

Base complementarity between adenine and thymine and between guanine and cytosine causes certain constancies concerning the base composition of double-stranded DNA:

1. The amount of adenine equals the amount of thymine.
2. The amount of guanine equals the amount of cytosine.
3. The total amount of purines (A+G) equals the total amount of pyrimidines (C+T).

The relationships between the quantity of bases contained in DNA were first recognized by Erwin Chargaff and are sometimes called "Chargaff's rules." They result from the fact that A base pairs with T and G base pairs with C in double-stranded DNA. In single-stranded DNA, however, the bases are not paired. Thus, Chargaff's rules do not apply. (Single-stranded DNAs *can*, in fact, fold upon themselves forming hydrogen bonded base pairs over very short regions. Thus, "single-stranded" DNAs often adopt a conformation that includes both single- and double-stranded regions. The double-stranded regions, however, are few in number and short in length. They do not significantly affect the base composition of single-stranded DNAs.)

Which of the above DNAs follow Chargaff's rules? In phages P4, C2, PZA, and Cp-1, all of Chargaff's rules apply. Such DNAs are double-stranded. The remaining phages (Pf3, IKe, S13, and fd) are single-stranded.

---

5. The DNA of many organisms contains a substantial fraction of modified or unusual bases. In human DNA, for example, approximately 5% of cytosine residues are modified to 5-methylcytosine after DNA replication. Human DNA contains 31% thymine. What percent of human DNA is 5-methylcytosine?

Answer Approximately 0.95%

6. If all the DNA of the *Zea mays* (corn) haploid genome were stretched out "end-to-end" as B-form DNA, the resulting "molecule" would be 1.8 meters long! How many picograms of DNA does this represent? Adjacent base pairs in B-form DNA are separated by 0.34 nanometers. The molecular weight of an average base pair of DNA is 660 g/mole. Some conversions you might need to answer this question are shown below.

$$1 \text{ picogram (pg)} = 10^{-12} \text{ grams (g)}$$
$$1 \text{ nanometer (nm)} = 10^{-9} \text{ meters (m)}$$
$$\text{Avogadro's number} = 6 \times 10^{23}$$

➤ ***Solution***

Answer 5.8 picograms

Converting meters of DNA into picograms of DNA requires careful interconversion of differing units of measurement. The question states that the molecular weight of an average base pair is 660 g/mole. If we first calculate the number of base pairs in 1.8 meters of DNA, we can then calculate the molecular weight of the genome. The molecular weight of the genome represents the mass of a mole of the genome. To determine the mass of a single molecule of the genome, we divide the molecular weight by Avogadro's number. The following three calculations perform these interconversions:

1. Calculate the number of base pairs (bp) contained in 1.8 meters of DNA:

$$1.8 \text{ m} \times 10^9 \text{ nm/m}$$
$$= 1.8 \times 10^9 \text{ nm of DNA in the genome}$$
$$1.8 \times 10^9 \text{ nm} \div 0.34 \text{ nm/bp}$$
$$= 5.3 \times 10^9 \text{ bp of DNA in the genome}$$

2. Calculate the molecular weight of the genome:

$$5.3 \times 10^9 \text{ bp} \times 660 \text{ g/mole/bp}$$
$$= 3.5 \times 10^{12} \text{ g/mole}$$

3. Divide by Avogadro's number to calculate the weight of a single molecule:

$$3.5 \times 10^{12} \text{ g/mole} \div 6 \times 10^{23} \text{ molecules/mole}$$
$$= 5.8 \times 10^{-12} \text{ g/molecule}$$

Thus, the maize genome consists of 5.8 picograms of DNA.

7. Depending on its sequence and on the methods of its isolation, double-stranded DNA can exist in at least three different structural conformations (A-form, B-form, and Z-form). DNA adopts the B-form in living cells, but it can adopt the alternative structures under special circumstances. Which of the following physical properties of DNA are *different* in A-form and B-form DNA? Check all that apply.

    ____ The helical pitch (base pairs per helical turn)
    ____ The helix direction (right-handed or left-handed helix)
    ____ The inclination (tilt) of bases away from the helix axis
    ____ The width and depth of the major and minor grooves
    ____ The overall width of the double helix (in the direction perpendicular to the helix axis)

Answer

    _X_ The helical pitch (base pairs per helical turn)
    ____ The helix direction (right-handed or left-handed helix)
    _X_ The inclination (tilt) of bases away from the helix axis
    _X_ The width and depth of the major and minor grooves
    _X_ The overall width of the double helix (in the direction perpendicular to the helix axis)

8. In certain bacteriophage, modified or unusual pyrimidines replace thymine or cytosine in double-stranded DNA. In the *Bacillus subtilis* phage SPO1, for example, all thymine is replaced in phage DNA with 5-hydroxymethyl uracil (HmUra). HmUra base pairs with adenine in SPO1 DNA. If the base composition of SPO1 DNA were 28% cytosine, what proportion is 5-hydroxymethyl uracil?

➤ *Solution*

Answer: 22%

Because SPO1 DNA is double-stranded, Chargaff's rules apply, with the modification that hydroxymethyl uracil replaces thymine. Chargaff's rules state:

1. The amount of adenine equals the amount of thymine.
2. The amount of guanine equals the amount of cytosine.
3. The total amount of purines (A+G) equals the total amount of pyrimidines (C+T).

If 28% of the bases are cytosine, then 28% of the bases are guanine. This totals 56%. The remaining bases (44%) are equal quantities of adenine and HmUra. Thus, 22% of SPO1 DNA is HmUra.

9. Bacteriophage ΦX174 contains single-stranded DNA within the phage particle. After being injected into a host cell, the single-stranded ΦX174 DNA is converted to a double-stranded replication intermediate. Replication of the double-stranded intermediate produces single-stranded products, which are packaged into phage heads. You are given two samples of ΦX174 DNA. Your analysis indicates that sample #1 contains 24.0% adenine and 21.3% cytosine, and that sample #2 contains 27.7% thymine and 22.3% guanine. Which of these samples contains single-stranded DNA? Which contains double-stranded replication-form DNA?

Answer

Sample #1 contains single-stranded ΦX174 DNA. Sample #2 contains double-stranded ΦX174 DNA.

10. The sequence of a short strand of DNA is shown below.

5′-CATCGACATTGCGAGC-3′

What is the sequence of an RNA strand complementary to this DNA? Indicate the 5′ and 3′ termini of the RNA.

➤ *Solution*

Answer: 5′-GCUCGCAAUGUCGAUG-3′

It is important to consider three structural features of DNA and RNA when answering this question:

1. RNA contains uracil in place of thymine.
2. G base pairs with C in both DNA and RNA; A base pairs with T in DNA and U in RNA.
3. The two strands of double-stranded nucleic acids are antiparallel.

Nucleotide sequences can either be written 5′-to-3′ (left-to-right) or 3′-to-5′. Both answers are equivalent, but sequences of single-stranded DNA or RNA are more commonly written with 5′ on the left and 3′ on the right. Thus, the 5′ (left) end of the complementary RNA must hydrogen bond with the 3′ (right) end of the DNA sequence shown above. And, the 3′ (right) end of the complementary RNA must hydrogen bond with the 5′(left) end of the DNA sequence shown above. The RNA complement of the above sequence is, therefore,

5′-GCUCGCAAUGUCGAUG-3′

11. The two strands of double-stranded DNA must be unwound prior to their replication by the enzyme DNA helicase. In *E. coli*, a replication fork moves forward at the rate of 500 bas e pairs per second. How many turns of the double helix must be removed from the template DNA in 1 minute for each replication fork?

Answer

3,000 complete turns of the double helix are removed each minute!

12. A culture of *E. coli* growing in normal medium is transferred into medium containing a heavy isotope of nitrogen (^{15}N) and grown for two additional generations. DNA is then harvested from the culture and analyzed by equilibrium density gradient centrifugation to determine its buoyant density. What proportion of the DNA will have both strands labeled (heavy/heavy), one strand labeled (heavy/light), and no strands labeled (light/light)?

➤ *Solution*

Answer: One half will be heavy/heavy; one half will be heavy/light.

The above experiment is similar to that of Meselson and Stahl. Their famous experiment demonstrated that DNA replication is semiconservative, meaning that parental DNA strands remain intact following DNA replication, with each daughter molecule inheriting one parental strand and one newly synthesized strand. Parental DNA strands are the template upon which newly-synthesized strands are assembled.

Prior to transfer of the cells into density-labeled media, all of the DNA is unlabeled (light/light). Following transfer into ^{15}N-containing medium and one cycle of DNA replication, each newly-synthesized strand is labeled, while each parental (template) strand is unlabeled. Thus, all of the DNA is heavy/light. These heavy/light DNAs constitute the parental strands for the next cycle of DNA replication. After a second round of DNA replication, both the heavy and light parental strands are base paired with a newly-synthesized heavy strand. Thus after two generations of growth, half of the DNA is heavy/heavy and half is heavy/light.

---

13. On the lagging strand of DNA synthesis, what is the relative order in which the following enzymes function during synthesis of a single Okazaki fragment and its subsequent incorporation into high molecular weight DNA?

1. DNA polymerase III
2. DNA primase
3. DNA ligase
4. DNA helicase

Answer

DNA helicase, followed by primase, followed by polymerase, followed by ligase

14. *E. coli* contains at least three different DNA polymerases. One of these, DNA polymerase III, is primarily involved in chromosomal DNA replication. DNA polymerase III contains three different enzymatic activities: a $5' \rightarrow 3'$ polymerase, a $5' \rightarrow 3'$ exonuclease, and a $3' \rightarrow 5'$ exonuclease. Consider a mutant DNA polymerase III in which the $3' \rightarrow 5'$ exonuclease activity is destroyed, but the other activities remain unaffected. What effect is this mutation likely to have on the rate and the fidelity of DNA chain elongation *in vivo*?

➤ *Solution*

Answer: The rate of replication will not be significantly affected. The rate of replication errors (base substitutions during replication) will be increased substantially.

The $3' \rightarrow 5'$ exonuclease activity of DNA polymerase is often termed its "proofreading" or "editing" function. This enzymatic activity only comes into play when DNA polymerase makes an error while inserting nucleotides into a growing chain. The $5' \rightarrow 3'$ polymerase activity of DNA polymerase is remarkably faithful. By using Watson-Crick rules of base pairing, DNA polymerase inserts a correct nucleotide approximately 99.99% of the time. In approximately 1 per 10,000 nucleotide additions, however, DNA polymerase inserts an incorrect nucleotide. Such errors, if unrepaired, lead to mutations in DNA.

The $3' \rightarrow 5'$ exonuclease of DNA polymerase repairs about 99% of such base misincorporation errors. After each cycle of nucleotide addition, DNA polymerase "checks" to see if the nucleotide inserted in the previous cycle is base paired properly with its template. If it is, DNA polymerase proceeds to the next cycle of nucleotide addition. If it is not base paired properly, the $3' \rightarrow 5'$ exonuclease removes the 3'-terminal nucleotide before beginning another cycle of nucleotide addition. Thus, DNA polymerase checks the accuracy of its replication at each cycle. (The *E. coli* enzyme does this at approximately 500 nucleotide additions per second!) A mutant DNA polymerase in which the $3' \rightarrow 5'$ exonuclease is nonfunctional will be unable to correct its polymerization errors. Thus, the frequency of base substitution mutations will increase dramatically. Because proofreading occurs so infrequently, however, the overall rate of chain elongation by the mutant polymerase will not be significantly affected.

15. A region of DNA containing a single-stranded gap is shown in the diagram on p. 200. The 5'-to-3' polarity of only the upper strand is shown.

Is the terminal phosphate (P) of the bottom strand located at the 5′ or the 3′ end of the strand to which it is attached?

➤ **Solution**

Answer: The terminal phosphate is located at the 5′ end.

The two strands of double-stranded DNA are antiparallel. If the top strand of the diagram has its 5′ end on the right, both bottom strands have their 5′ ends on the left. Thus, the terminal phosphate of the bottom strand is a 5′ phosphate.

16. A culture of *E. coli* growing in normal medium is transferred into medium containing a heavy isotope of nitrogen (^{15}N) and grown for one additional generation. DNA is harvested from the culture and sheared into fragments that are on average 400 base pairs long. One half of the DNA (sample #1) is heated to 100°C and allowed to cool slowly. Sample #1 is then treated with S1 exonuclease, an enzyme that degrades all single-stranded DNA (releasing mononucleotides) but which does not affect double-stranded DNA. Sample #1 is then analyzed by equilibrium density gradient centrifugation to determine its buoyant density. The other half (sample #2) is analyzed by equilibrium density gradient centrifugation without prior heating or exonuclease treatment. What proportion of the DNA in each sample will have both strands labeled (heavy/heavy), one strand labeled (heavy/light), and no strands labeled (light/light)?

➤ **Solution**

Answer:

Sample #1: ¼ of the DNA is heavy/heavy; ½ is heavy/light; ¼ is light/light.

Sample #2: all of the DNA is heavy/light.

Replication of DNA is semiconservative, meaning that parental DNA strands remain intact following DNA replication, with each daughter molecule inheriting one parental strand and one newly synthesized strand. After one generation of growth in ^{15}N-containing medium, all of the DNA contains one parental (light) strand and one newly-synthesized (heavy) strand. Thus, it is all of heavy/light density.

The heavy/light DNA is then sheared into small fragments. Sample #2 is analyzed directly by equilibrium density gradient centrifugation. Its density will be all heavy/light. Sample #1 is first denatured and then allowed to reanneal. DNA fragments with base sequence complementarity base pair randomly. When two ^{15}N-labeled fragments reanneal, the result is heavy/heavy DNA. When labeled and unlabeled fragments reanneal, the result is heavy/light DNA. When two unlabeled fragments reanneal, the result is light/light DNA. Random association of single strands yields a 1:2:1 ratio of heavy/heavy, heavy/light, and light/light molecules, respectively. Subsequent treatment of the hybrids with S1 exonuclease removes any single-stranded regions at the ends of the molecules as well as any single strands that have not reannealed. (Such single strands complicate equilibrium density gradient centrifugation. Single-stranded DNA has a different buoyant density than double-stranded DNA.)

---

17. The double-stranded character of DNA is stabilized by weak noncovalent chemical interactions, the most prominent of which are the hydrogen bonds of A:T and G:C base pairs. Such weak interactions can be disrupted by elevated temperature. The temperature above which a sample of DNA is single-stranded rather than double-stranded is termed its melting temperature (T_m). Is the melting temperature of $G_{30}{:}C_{30}$ (double-stranded DNA containing 30 G residues on one strand and 30 C residues on the opposite strand) higher than, lower than, or equal to that of $A_{30}{:}T_{30}$?

➤ *Solution*

Answer: $G_{30}{:}C_{30}$ has a higher T_m than $A_{30}{:}T_{30}$.

Hydrogen bonds are an important noncovalent interaction that stabilizes double-stranded DNA. (They are not, however, the *only* stabilizing interaction. Hydrophobic base stacking interactions between adjacent base pairs also stabilizes the double-stranded character of DNA.) Hydrogen bonding between G:C base pairs, which involves three hydrogen bonds, is stronger than between A:T base pairs, which involves two hydrogen bonds. Double-stranded DNA denatures ("melts") only when most or all of the hydrogen bonds between the DNA strands are disrupted. Because G:C hydrogen bonding is stronger, DNAs rich in G:C base pairs denature at a higher temperature than DNAs rich in AT base pairs.

---

18. The substrates and products of four enzymes are shown in the diagram below. What enzyme catalyzes each reaction?

Substrate   **Product**

A.   5'-CGATGGAC-3'   5'-CGATGGACTAGGCATT-3'
      3'-GCTACCTGATCCGTAA-5'   3'-GCTACCTGATCCGTAA-5'

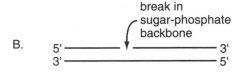

B.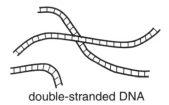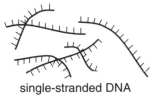

C.   5'-CTTGCATGGTAA-3'   5'-CTTGCATGGTA-3'
      3'-GAACGTACCATGCAAGGT-5'   3'-GAACGTACCATGCAAGGT-5'

D.

double-stranded DNA   single-stranded DNA

Answer
(a) DNA polymerase
(b) DNA ligase
(c) DNA polymerase
(d) DNA helicase

19. A substantial fraction of the DNA contained in a typical eukaryote is moderately repetitive. Such DNA usually consists of many different "families" of sequence elements. Any individual repetitive element is similar or identical in DNA sequence to other members of its family, but the sequence of one family is different from that of other families. In a eukaryote whose haploid genome is 10^8 base pairs, 10% of the DNA is moderately repetitive. The average size of a segment of moderately repetitive DNA is 1,000 base pairs. The average family consists of 50 members dispersed throughout the genome. How many different families of moderately repetitive DNA are present in this organism?

➤ *Solution*

Answer: 200 families

If the genome is 10^8 base pairs long, and if 10% of the genome consists of moderately repetitive DNA, then 10^7 base pairs of moderately repetitive DNA are present throughout the genome. The average size of a segment of moderately repetitive DNA is 10^3 base pairs. Thus $10^7 \div 10^3 = 10^4$ repeated sequences of 1,000 base pairs each are dispersed throughout the genome. These sequences are related as families, with each family consisting on average of 50 members. Thus, $10^4 \div 50 = 200$ different families of moderately repetitive DNA are present in the genome.

20. A fragment of double-stranded DNA (~3,000 base pairs) is heated to 100°C to denature the strands and cooled slowly (allowing strands to reanneal). The resulting DNA molecules are visualized in an electron microscope. Molecules shown in the diagram below are observed.

—— double-stranded DNA

– – – single-stranded DNA

Describe the structure of the initial fragment of double-stranded DNA. How does the DNA sequence at its left end compare to the DNA sequence at its right end?

Answer

The double-stranded DNA contains an inverted repeat at its termini. That is, the sequence at one end of the molecule (reading 5′-to-3′ on one strand) is identical to the sequence at the other end of the molecule (reading 5′-to-3′ on the opposite strand).

21. The sequences of five DNA or RNA oligonucleotides are shown below.

> Oligonucleotide #1: 5′-CATAGCACCG-3′
> Oligonucleotide #2: 5′-GTATCGTGGC-3′
> Oligonucleotide #3: 5′-GUAUCGUGGC-3′
> Oligonucleotide #4: 5′-GCCACGATAC-3′
> Oligonucleotide #5: 5′-CGGUGCUAUG-3′

Oligonucleotides #1-#5 are mixed together, heated to 100°C, and cooled slowly (allowing single strands to anneal). Draw the sequences of all double-stranded nucleic acids that result.

Answer

Three different double-stranded nucleic acids result. Oligonucleotides #1 and #5 form DNA/RNA hybrids of sequence:

> 5′-CATAGCACCG-3′
> 3′-GUAUCGUGGC-5′

Oligonucleotides #2 and #4 form DNA/DNA hybrids of sequence:

> 5′-GTATCGTGGC-3′
> 3′-CATAGCACCG-5′

Oligonucleotides #3 and #4 form DNA/RNA hybrids of sequence:

> 5′-GUAUCGUGGC-3′
> 3′-CATAGCACCG-5′

22. Consider two fragments of double-stranded DNA: fragment #1 is 3 kilobases (3,000 base pairs) long; fragment #2 is identical to fragment #1 except that it contains an inversion of the central 1 kilobase segment. Fragments #1 and #2 are mixed together, heated to 100°C to denature the strands and cooled slowly (allowing strands to reanneal). The resulting DNA molecules are visualized in an electron microscope. Draw three different structures expected. Indicate double-stranded and single-stranded regions of the molecules.

Answer

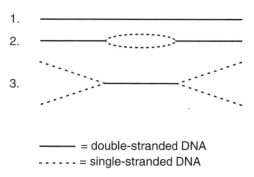

```
          ——— = double-stranded DNA
          · · · · · · = single-stranded DNA
```

23. You have isolated three deletion mutations affecting your favorite bacteriophage. The relative size and location of these deletions is shown on the diagram below.

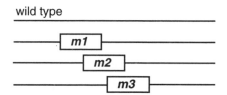

DNAs of wild type (WT) and *m1* through *m3* are mixed together in pairs, heated to denature the strands, and allowed to reanneal. The DNAs are then examined in an electron microscope under conditions in which single-stranded and double-stranded regions can be distinguished. Draw the structures expected of each heteroduplex molecule (WT:*m1*, WT:*m2*, *m1*:*m2*, etc.).

Answer

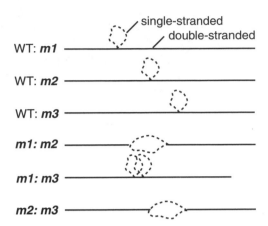

24. A Holliday intermediate, formed during homologous recombination, is shown in Part A of the diagram below.

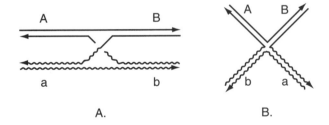

A.   B.

Alleles *A,a* and *B,b* of two genes that flank the site of crossing over are indicated. Arrowheads represent the 3′ termini of DNA strands. Is the DNA structure shown in Part B of the diagram topologically equivalent to the Holliday junction of Part A?

Answer Yes

25. Yeast genes *A,a* and *B,b* are separated by approximately 3 map units. A yeast of genotype *AB/ab* is induced to undergo meiosis. One hundred asci (tetrads) are dissected, and the genotypes of individual spores are determined. As shown in the table on p. 207, four different classes of asci are observed.

Ascus class	Genotype of Spores	Number
1	1 AB, 1 ab, 1 Ab, 1 aB	2
2	1 AB, 2 ab, 1 Ab	1
3	2 AB, 1 Ab, 1 ab	1
4	2 AB, 2 ab	96

Which of these ascus classes results from gene conversion during meiosis?

➤ **Solution**

Answer: Ascus classes 2 and 3 result from gene conversion.

In most meioses, alleles exhibit 2:2 segregation. Class 4 asci, for example, contain 2 spores of genotype *AB* and 2 of genotype *ab*. Such spores are nonrecombinant, but more importantly for this question, there are 2 copies each of alleles *A* and *a*, and 2 copies each of alleles *B* and *b*. Each allele pair segregates 2:2. Such segregation follows from the orderly disjunction of homologues during meiosis I.

Gene conversion, on the other hand, results in a 3:1 segregation of alleles. For a gene that undergoes gene conversion, 2 copies of each allele *enter* meiosis, but 3 copies of 1 allele and 1 copy of the other allele *exit* meiosis. Gene conversion "converts" one allele into another during meiosis. Which asci exhibit 3:1 segregation of either the *A,a* or *B,b* genes? Ascus class 2 contains 3 *b* alleles and 1 *B* allele. Similarly, ascus class 3 contains 3 *A* alleles and 1 *a* allele. Class 2 and 3 asci result from gene conversions affecting the *B,b* and *A,a* genes, respectively. Ascus class 1 exhibits 2:2 segregation; it did not result from gene conversion. Class 1 asci are tetratypes, resulting from a single reciprocal crossover between the *A,a* and *B,b* genes.

26. *Arg1-1* and *arg1-2* are alleles of the same arginine biosynthetic gene. Both *arg1-1* and *arg1-2* mutants are arginine auxotrophs. An Arg⁻ diploid of genotype *arg1-1+/+arg1-2* is induced to undergo meiosis. Two hundred asci (tetrads) are dissected, and the genotypes of their spores are determined: 199 contain 4 Arg⁻:0 Arg⁺ spores, and 1 contains 3 Arg⁻:1 Arg⁺ spores. Did the exceptional ascus result from gene conversion or from crossing over (reciprocal recombination)? If you have insufficient information to be certain, what additional information is needed?

➤ **Solution**

Answer: It cannot be determined whether the exceptional ascus resulted from gene conversion or from crossing over. The genotypes of the three Arg⁻ spores are needed.

The exceptional ascus contains one Arg⁺ spore, which must have resulted either from gene conversion or from crossing over (reciprocal recombination). The origins of the Arg⁺ spore (gene conversion or crossing over) affects the genotypes of the three Arg⁻ spores. If the Arg⁺ spore resulted from reciprocal recombination, then one of the Arg⁻ spores is a double mutant of genotype *arg1-1 arg1-2*, while the other two are *arg1-1* and *arg1-2* single mutants. If the Arg⁺ spore resulted from gene conversion, then all three of the Arg⁻ spores are either genotype *arg1-1* or *arg1-2*. Thus, the genotypes of the Arg⁻ spores must be known before the origins of the Arg⁺ spore can be established.

27. Genes *a* and *b* of *Neurospora* are linked to each other and to their centromere. A diploid of genotype *a+/+b* is put through meiosis and genotypes of the resulting ascospores are determined. Many different ascus patterns are observed, among them being:

	GENOTYPE OF ASCOSPORES							
Ascus type 1	+b	+b	+b	+b	a+	a+	a+	a+
Ascus type 2	ab	ab	a+	a+	++	++	+b	+b
Ascus type 3	a+	a+	++	++	+b	+b	+b	+b
Ascus type 4	a+	a+	+b	+b	a+	a+	+b	+b
Ascus type 5	+b	+b	++	++	ab	ab	a+	a+
Ascus type 6	++	++	ab	ab	a+	a+	ab	ab
Ascus type 7	+b	+b	+b	+b	a+	a+	ab	a+

Note that not all ascus patterns are equally frequent. Some are common; some are rare. Which of these asci result from gene conversion during meiosis of the diploid?

Answer Ascus types 3, 6, and 7

28. In many organisms, a substantial fraction of cytosine contained in DNA is enzymatically modified after DNA replication to 5-methylcytosine. Both cytosine and 5-methylcytosine base pair with guanine. As shown in the diagram on p. 209, spontaneous deamination of 5-methylcytosine yields thymine.

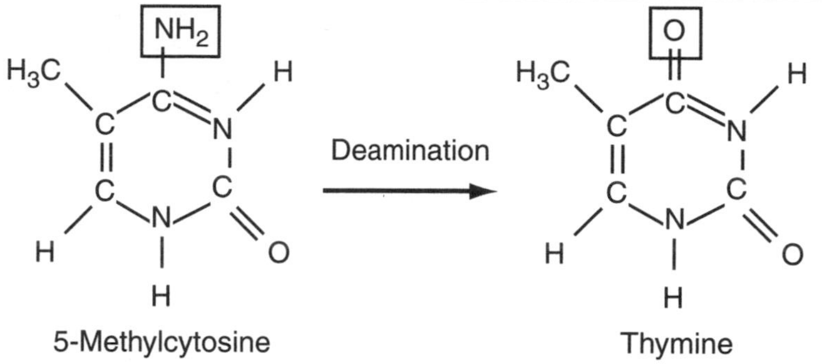

Deamination

5-Methylcytosine   Thymine

The DNA sequence of a single strand of a portion of a wild-type gene and that of five spontaneous mutant alleles is shown below.

    Wild type: 5′-ACTCAGTCGGCCGCCACG-3′

    Allele #1: 5′-ACTCA**C**TCGGCCGCCACG-3′

    Allele #2: 5′-ACTCAGT**T**GGCCGCCACG-3′

    Allele #3: 5′-ACTCAGTCGGCCGCC**G**CG-3′

    Allele #4: 5′-ACTCAG**A**CGGCCGCCACG-3′

    Allele #5: 5′-ACTCAGTCGGCC**A**CCACG-3′

Which of these alleles might have arisen following spontaneous deamination of 5-methylcytosine?

➤ *Solution*

Answer: Alleles #2 and #5

Each of the mutant alleles contains a single nucleotide substitution relative to the wild-type sequence. In allele #1, for example, a G of the wild-type sequence is replaced with a C. Is this substitution consistent with deamination of 5-methylcytosine? To answer this, we must examine the consequences of such deamination.

During DNA replication, a base pair between G and 5-methylcytosine (abbreviated G:mC) behaves exactly like a base pair between G and C (abbreviated G:C). The G base pairs with C, and the mC base pairs with G during replication. Thus, the immediate products of DNA replication are a G:C and a mC:G base pair. (The C residues of such G:C base pairs are subsequently methylated, thereby regenerating G:mC base pairs.) What happens if the 5-methylcytosine residue is deaminated prior to DNA replication? Deamination produces a G:T base pair. During DNA replication the G base pairs with C and the T base pairs with A. Thus, the products of DNA replication are one G:C and one A:T base pair. The molecule containing a G:C base pair is wild-type; the daughter containing an A:T base pair is mutant. Thus, spontaneous deamination of 5-methylcytosine changes G:C base pairs into A:T base pairs. Which of the mutant alleles contain G→A or C→T substitutions? Alleles #2 and #5.

---

29. The mutagen 5-bromouracil (5-BU) is a base analog of thymine. When added to the growth medium of cells, 5-BU is assimilated into nucleotide triphosphates and incorporated by DNA polymerase into DNA. 5-BU usually base pairs as does thymine, but it occasionally base pairs as does cytosine. The DNA sequence of a single strand of a portion of a wild-type gene and that of five mutant alleles is shown below.

Wild type: 5'-GCACGGTCATGCGCTCAGG-3'
Allele #1: 5'-GCAC**C**GTCATGCGCTCAGG-3'
Allele #2: 5'-GCACGGTCATG**T**GCTCAGG-3'
Allele #3: 5'-GC**G**CGGTCATGCGCTCAGG-3'
Allele #4: 5'-GCACGGTCATGCGC**C**CAGG-3'
Allele #5: 5'-GCACGGTC**T**TGCGCTCAGG-3'

Which of these alleles might have been induced by 5-BU?

Answer Alleles #2, #3, and #4

30. When purified DNA is treated with nitrous acid (NA), deamination of cytosine and adenine produces uracil and hypoxanthine, respectively. Uracil pairs as does thymine; hypoxanthine pairs as does guanine. The DNA sequence of a single strand of a portion of a wild-type gene and that of five mutant alleles is shown below.

Wild type: 5'-GATCTCGGACCGTCATGCCA-3'
Allele #1: 5'-GATCTCGG**G**CCGTCATGCCA-3'
Allele #2: 5'-GATCTCGGA**G**CGTCATGCCA-3'
Allele #3: 5'-GATCTCG**A**ACCGTCATGCCA-3'
Allele #4: 5'-GATCTCGGACCG**A**CATGCCA-3'
Allele #5: 5'-GATCTCG**C**ACCGTCATGCCA-3'

Which of these alleles might have arisen following treatment of plasmid DNA with nitrous acid and transformation of recipient cells with the treated DNA?

Answer Alleles #1 and #3.

31. The substrates and products of four enzymes involved in DNA repair are shown in the diagram below. Arrowheads in parts A and D indicate 3′ termini of DNA strands.

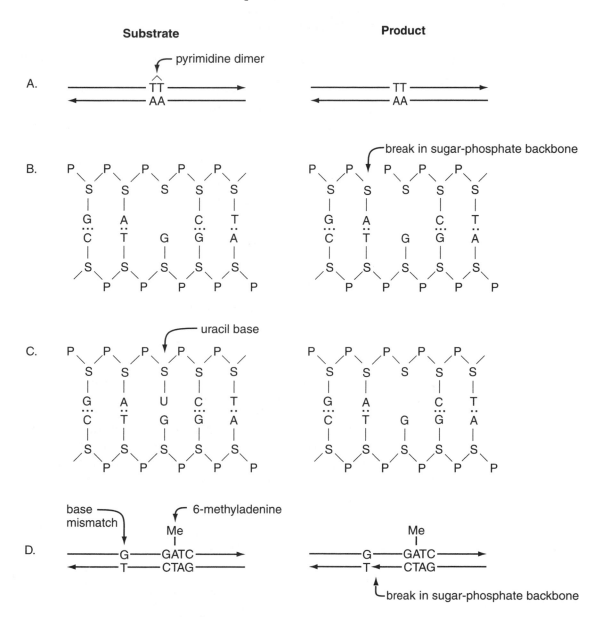

What enzyme catalyzes each reaction?

Answer

(a) DNA photolyase
(b) AP endonuclease
(c) uracil-DNA glycosylase
(d) *mutH* endonuclease

32. The extraordinary accuracy with which a genome is replicated is due in part to many different systems of DNA repair. Such systems reverse the effects of spontaneous and induced DNA damage. Nucleotide excision repair and DNA base excision repair are two different systems of DNA repair. Which of the following enzymes function in both nucleotide excision repair and base excision repair? Check all that apply.

___ DNA helicase
___ AP endonuclease
___ 5′ → 3′ exonuclease
___ DNA ligase
___ DNA glycosylase

Answer

___ DNA helicase
___ AP endonuclease
___ 5′ → 3′ exonuclease
_X_ DNA ligase
___ DNA glycosylase

33. Many organisms contain modified nucleotides within their DNA. In vertebrates, for example, a substantial fraction of cytosine is enzymatically modified to 5-methylcytosine. In *E. coli*, adenine residues that are located within the tetranucleotide sequence 5′-GATC-3′ are enzymatically modified to 6-methyladenine. In which system(s) of DNA repair does 6-methyladenine perform important functions in *E. coli*? Check all that apply.

___ Nucleotide excision repair
___ DNA base excision repair
___ Mismatch repair
___ Photoreactivation

Answer

___ Nucleotide excision repair
___ DNA base excision repair
_X_ Mismatch repair
___ Photoreactivation

Gene Expression

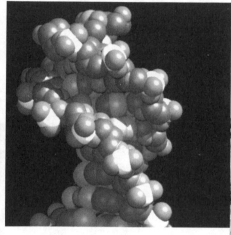

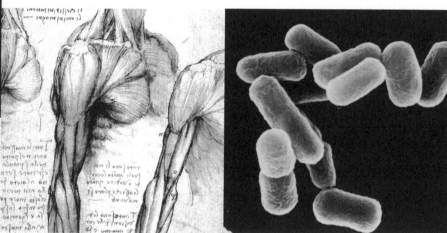

Summary _____

DNA encodes the primary amino acid sequence of proteins and directions for when and how much of a particular protein to produce. DNA is **transcribed** to make **messenger RNA (mRNA)**, which is then **translated** to make protein. Together, transcription and translation constitute **gene expression.** DNA content is constant in most cells, so all the genes of an organism are present in the same relative proportions. The amount of a specific mRNA in a given cell type and the rate at which this mRNA is translated control the abundance of the corresponding protein. For example, silk worms produce large amounts of fibroin mRNA and protein when they pupate but not as adult moths; only immunologically stimulated B cells produce immunoglobulin mRNA and proteins to fight off viral infections; and only developing endosperm cells of corn kernels accumulate the enzymes required for starch synthesis.

There are similarities and differences in the ways gene expression occurs in prokaryotes and eukaryotes. The **promoter** is a DNA sequence which serves as a binding site for **RNA polymerase**, and is found 5′ to the coding portion of most genes. In bacteria, multiple protein-coding regions are often transcribed from a single promoter to yield a **polycistronic** mRNA. The multiple genes expressed from a single promoter are called an **operon.** For example, the *lac Z, Y,* and *A* genes are transcribed as a single mRNA, and their protein products work together to allow bacteria to use lactose as a carbon source. The organization of many genes in bacteria as operons allows for **coordinate control** of gene expression.

As RNA polymerase binds to the promoter, a short region of the two strands of the DNA denatures to form an **open promoter complex**. Transcription, like DNA replication, occurs in a 5′ to 3′ direction. Complementary **ribonucleotides** hydrogen bond to the template strand and are added by RNA polymerase to the 3′ end of the elongating RNA molecule. RNA contains the pyrimidine base **uracil** in place of thymine as in DNA. Transcription in bacteria continues until RNA polymerase encounters specific termination signals.

Transcription and translation are coupled in bacteria; once RNA polymerase has transcribed several hundred bases, translation begins. Bacterial ribosomes initiate translation near the 5′ end of a mRNA. Initiation of translation is facilitated by base pairing between a short region at the 3′ end of the 16S (small subunit) ribosomal RNA and a complementary region near the 5' end of a mRNA (the **Shine-Dalgarno sequence**). Binding of the small ribosomal subunit to the mRNA leads to assembly of the **translation initiation complex**, consisting of initiation factors, the message, GTP, and the small ribosomal subunit. The first aminoacyl-tRNA which binds is N-formyl-methionyl-tRNA (fMet-tRNA^{Met}). Translation elongation begins following joining of the initiation complex by the ribosomal large subunit, hydrolysis of GTP, and dissociation of the initiation factors.

Transfer RNAs (tRNAs) work with ribosomes to "decode" mRNAs. This decoding process converts the linear sequence of nucleotides in a mRNA to a linear sequence of amino acids in a protein. Transfer RNAs are small, compact "adaptors" that play a key role in converting nucleotide sequences to amino acid sequences. Transfer RNAs in solution adopt a "folded cloverleaf" structure that resembles the shape of an "L." Amino acids are "activated" for translation by covalently attaching them to the 3' ends of specific tRNAs. This tRNA "charging" reaction is catalyzed by **aminoacyl-tRNA synthetases**. There are twenty different aminoacyl-tRNA synthetases, one for each amino acid. Each aminoacyl-tRNA synthetase attaches its amino acid only to appropriate tRNAs. Each three-base triplet in mRNA (a **codon**) specifies one of twenty different amino acids by its ability to base pair with the **anticodon** region of a specific tRNA. The accuracy and fidelity of translation is determined, in part, by the specificity of the codon:anticodon base pairing. Because some amino acids are encoded by several different codons, the triplet code is said to be **degenerate**. With one exception, base pairing between a codon and anticodon follows standard Watson-Crick rules. The exception concerns base pairing between nucleotides located at the 3' end of the codon and the 5' end of the anticodon (the **wobble** position). A wider variety of base pairing (for example, G:U pairing) is allowed at the wobble position than elsewhere. The wobble position allows some tRNAs to base pair with several different codons differing from each other only by their third (3') nucleotide.

During translation **elongation**, an orderly sequence of events adds amino acids sequentially to the carboxyl terminus of the growing polypeptide chain. Interactions between aminoacyl-tRNAs, mRNA, and ribosomes are instrumental to the fidelity of this process. Three different sites on a ribosome interact with tRNAs during a cycle of elongation. A **peptidyl-tRNA**, bound at the **P site**, is covalently attached at its 3' end to the carboxyl-terminal amino acid of the growing polypeptide chain. The next amino acid to be added enters the ribosome as an aminoacyl-tRNA bound at the **A site**. Such aminoacyl-tRNAs are delivered to the A site by translation **elongation factors**. Correct codon:anticodon base pairing is necessary to stabilize an aminoacyl-tRNA at the A site. Occupancy of the A site positions an aminoacyl-tRNA in preparation for **peptide bond** formation. When both the P site and A site are occupied, the **peptidyl transferase** activity of the ribosome forms a peptide bond between the carboxyl terminus of the peptidyl-tRNA peptide and the amino group of the amino acid bound as an aminoacyl-tRNA at the A site. Thus, the growing peptide is transferred from the tRNA bound at the P site to the tRNA bound at the A site. The P site tRNA, which now does not contain an attached peptide, then exits the ribosome as a deacylated tRNA, interacting briefly during release with the ribosome **E site**. Following release of tRNA from the P site, the mRNA **translocates** by three bases relative to the ribosome, thereby shifting the A site peptidyl-tRNA into the P site and freeing the A site for a subsequent round of translation elongation. The polypeptide chain grows from the amino terminus towards the carboxyl terminus as the ribosome translocates from the 5′ to the 3′ end of the mRNA.

The third and final phase of translation occurs when ribosomes encounter one of three codons that do not have a corresponding aminoacyl-tRNA. Such codons are referred to as **nonsense** or **stop** codons. When a ribosome encounters a stop codon, **release factors** bind to the A site, converting the peptidyl transferase to a hydrolase. The ribosome then releases the polypeptide, thereby completing the process of translation. This is the **termination** phase of translation.

Eukaryotic gene expression differs from prokaryotic gene expression in several ways. Eukaryotes have three different classes of nuclear RNA polymerase that transcribe distinct classes of genes. RNA polymerase I transcribes ribosomal RNA genes. RNA polymerase II transcribes all genes that encode proteins (the vast majority of genes). Most nuclear genes of eukaryotes are **monocistronic**, meaning that they encode a single polypeptide. RNA polymerase III transcribes tRNA and 5S RNA genes.

Each class of eukaryotic polymerase interacts with its own set of transcription factors (TFIA, TFIIA, TFIIIA, etc.). Transcription factors bind to eukaryotic promoters, and RNA polymerase then binds to this protein/DNA complex. The TATA sequence element of Class II promoters is typically found 25 to 30 bp 5′ to the transcription initiation site. This *cis*-acting sequence element interacts with the basal transcription apparatus to position correctly the start site of transcription. Other *cis*-acting regulatory sites of eukaryotic promoters interact with sequence-specific DNA-binding transcription factors to either stimulate or repress nearby initiation of transcription.

The primary transcripts of most eukaryotic genes are **pre-mRNAs**, which are then extensively processed in the nucleus prior to being exported to the cytoplasm for translation as mature mRNAs. Most pre-mRNAs are matured by three distinct processing steps: **capping**, **splicing**, and **polyadenylation**. First, a **cap** consisting of a modified guanine nucleotide is added to the 5′ end of the mRNA using a 5′-to-5′ triphosphate linkage. Efficient translation of eukaryotic mRNAs requires a 5′ cap. The cap may also serve to protect mRNAs from degradation at their 5′ ends by ribonucleases. Second, **introns** are removed from pre-mRNAs by splicing. The coding regions of most eukaryotic genes are not contiguous in the DNA. Rather, protein-coding regions are interrupted by sequences that must be removed from pre-mRNAs before those transcripts can be translated. Splicing of a pre-mRNA occurs within large protein/RNA complexes while the pre-mRNA is being transcribed. Third, the 3′ end of mature mRNAs is created by cleavage of pre-mRNAs immediately downstream of a specific polyadenylation signal, and several dozen to several hundred adenosine residues are then added to the 3′ end of the mature mRNA. Such **poly(A) tails** may protect the 3′ ends of mRNAs from degradation by ribonucleases and modulate the efficiency of eukaryotic mRNA translation.

Self-Testing Questions

1. A point mutation in the *trpA* gene changes a UAC codon (tyrosine) to a UAA codon. The resulting mutant bacterial strain lacks tryptophan synthetase activity. Explain this result.

   Answer

   In the universal genetic code, there are three stop or nonsense codons. UAA is one such stop codon. This *trpA* mutation produces a truncated and apparently nonfunctional tryptophan synthase.

2. Compare the activities of *E. coli* DNA Polymerase III and RNA polymerase by describing the following properties of each:

	Pol III	RNA polymerase
(a) substrate(s)	_____	_____
(b) primers and templates	_____	_____
(c) products	_____	_____
(d) direction of polymerization	_____	_____

Answer

	Pol III	RNA polymerase
(a) substrate(s)	dNTPs (deoxyribo- nucleotide triphosphates)	NTPs (ribonucleotide triphosphates)
(b) primers and templates	RNA primer, both strands of DNA are used as template	no primers, only one strand of DNA is used as template
(c) products	double stranded DNA	RNA
(d) direction of polymerization	5′ to 3′	5′ to 3′

3. Identify the genes transcribed by each type of eukaryotic RNA polymerase found in the nuclei of eukaryotes.

Answer
Polymerase I: ribosomal RNA genes
Polymerase II: protein encoding genes, certain small RNAs
Polymerase III: tRNAs, 5S RNA; certain small RNAs

4. A tRNA has the anticodon 5′-UAC-3′. With which of the following codons is this anticodon complementary?
(a) 5′-CAU-3′
(b) 5′-AUG-3′
(c) 5′-GUA-3′
(d) 5′-UAC-3′

Answer

All nucleic acids base pair in an antiparallel fashion. Thus, the anticodon 5′-UAC-3′ is complementary to the codon 3′-AUG-5′. RNA and DNA sequences are conventionally written 5′-to-3′ (left-to-right). Thus, the codon 3′-AUG-5′ would be written as 5′-GUA-3′.

5. Fill in the blanks of the following sentences.
  (a) The cellular organelle at which translation takes place is termed the _____.
  (b) An anticodon of sequence 5′-GUA-3′ will base pair with a codon of sequence _____ (label the 5′ and 3′ termini).
  (c) Distant regulatory elements that control the expression of eukaryotic genes in *cis* are termed _____.
  (d) The modified guanine nucleotide found at the 5′ end of eukaryotic mRNAs is termed the _____.
  (e) Aminoacyl-tRNAs are delivered to the _____ site of the ribosome by elongation factor _____.
  (f) Transfer RNAs are "charged" with amino acids by the enzyme _____.
  (g) The three codons that signal translation termination in most organisms are _____, _____, and _____.
  (h) The principle of _____ explains why a single tRNA can often base pair with more than one codon.
  (i) The sequence of DNA to which RNA polymerase binds and within which transcription initiates is termed the

  _____.

  (j) Sequences removed by splicing from eukaryotic pre-mRNAs are termed _____.

Answers
  (a) The cellular organelle at which translation takes place is termed the <u>ribosome</u>.
  (b) An anticodon of sequence 5′-GUA-3′ will base pair with a codon of sequence <u>5′-UAC-3′</u>.
  (c) Distant regulatory elements that control the expression of eukaryotic genes in *cis* are termed <u>enhancers</u>.
  (d) The modified guanosine nucleotide found at the 5′ end of eukaryotic mRNAs is termed the <u>cap</u>.
  (e) Aminoacyl-tRNAs are delivered to the <u>P</u> site of the ribosome by elongation factor <u>Tu</u>.
  (f) Transfer RNAs are "charged" with amino acids by enzymes termed <u>aminoacyl-tRNA synthetases</u>.
  (g) The three codons that signal translation termination in most organisms are <u>UAG</u>, <u>UAA</u>, and <u>UGA</u>.
  (h) The principle of <u>wobble</u> explains why a single tRNA can often base pair with more than one codon.
  (i) The sequence of DNA to which RNA polymerase binds (directly in prokaryotes; indirectly in eukaryotes) and within which transcription initiates is termed the <u>promoter</u>.

(j) Sequences removed by splicing from eukaryotic pre-mRNAs are termed <u>introns</u>.

6. The following DNA sequence is transcribed into RNA, with the upper strand serving as the template strand.

5'-ATGCTACAGTC-3'
3'-TACGATGTCAG-5'

What is the sequence of the resulting RNA? Label the 5' and 3' termini.

Answer

5'-GACUGUAGCAU-3'

7. Midway through transcription of a particular mRNA molecule in bacteria, radioactive uracil is added to the growth medium, and synthesis of the RNA is completed. At which end (5' or 3') of the mRNA will the radiolabeled nucleotides be found? Midway through translation of a particular protein in bacteria, radioactive amino acids are added to the growth medium, and synthesis of the protein is completed. At which end of the protein (amino or carboxyl terminus) will the radiolabeled amino acids be found?

Answer

The mRNA will be radiolabeled at its 3' terminus. The protein will be radiolabeled at its carboxyl terminus.

8. When purified *E. coli* RNA polymerase holoenzyme (core enzyme plus sigma subunit) and ribonucleotide triphosphates are added to a promoter-containing bacterial DNA *in vitro*, transcription often initiates in a sequence-specific manner at the same site within the promoter as that initiated from the same promoter *in vivo*. Purified eukaryotic RNA polymerase II, however, is generally not capable of such sequence-specific initiation *in vitro*. What factors in addition to RNA polymerase II are needed for sequence-specific initiation of eukaryotic genes? How do they function during the transcription initiation process?

Answer

Eukaryotic RNA polymerase II does not directly bind to the promoter. Rather, basal and gene-specific transcription factors first bind to specific sites within the promoter. Only then does RNA polymerase II bind to such transcription initiation complexes.

9. The *lac* operon includes the genes *Z, Y,* and *A*. What effect does a mutation affecting the promoter of *lac Z* have on expression of *lac Y* and *lac A*?

Answer

lac Z, Y and *A* are transcribed from a single promoter, upstream of *lac Z*. A mutation of the promoter reducing *lac Z* transcription would also decrease transcription of *lac Y* and *A*. All three are coordinately controlled as an operon.

10. Purified cysteinyl-tRNA synthetase charges the amino acid cysteine on to tRNA^{Cys}, thereby producing cysteinyl-tRNA^{Cys} (an aminoacyl-tRNA). Purified cysteinyl-tRNA^{Cys} can be chemically altered *in vitro* such that the attached cysteine is converted to the amino acid alanine, thereby producing alanyl-tRNA^{Cys}. Alanyl-tRNA^{Cys} has the sequence of tRNA^{Cys} (including the anticodon), but it is charged with alanine instead of cysteine. Cysteinyl-tRNA^{Cys} or alanyl-tRNA^{Cys} are incubated with ribosomes in *in vitro* translation reactions that contain either poly(UGC) or poly(GCG) as artificial mRNAs. UGC is a cysteine codon; GCG is an alanine codon. Which of the following reactions are expected to synthesize polypeptides *in vitro*? What amino acid sequences will result?

	ADDED AMINOACYL-tRNA	
	cysteinyl-tRNA^{Cys}	alanyl-tRNA^{Cys}
Artificial mRNA		
poly(UGC)	_____	_____
poly(GCG)	_____	_____

Answer

	ADDED AMINOACYL-tRNA	
	cysteinyl-tRNA^{Cys}	alanyl-tRNA^{Cys}
Artificial mRNA		
poly(UGC)	poly-cysteine	poly-alanine
poly(GCG)	no polypeptide synthesis	no polypeptide synthesis

11. Initiation of translation in both prokaryotes and eukaryotes involves assembly of the translation initiation complex at AUG (or, occasionally, GUG) initiator codons. The initiation complex includes mRNA, the ribosomal small subunit, initiation factors, GTP, and fMet-tRNA^{Met} (prokaryotes) or tRNA^{Met} (eukaryotes). Methionine, therefore, is the first amino acid of most proteins. Is this methionine located at the amino terminus or the carboxyl terminus of the finished protein? Does translation initiation occur nearer the 5′ or 3′ end of the mRNA?

Answer

Methionine is the amino-terminal amino acid of most proteins (although it is occasionally removed from a mature protein after translation). Translation initiates near the 5′ end of mRNAs.

12. The DNA encoding most prokaryotic genes is colinear with its corresponding messenger RNA. Thus, if a prokaryotic mRNA is hybridized with a complementary single strand of the gene, the two molecules will be fully base paired throughout the length of the mRNA. This is often not the case in eukaryotes, however.

A purified eukaryotic mRNA is hybridized with a complementary single strand of the corresponding gene. The DNA strand includes the entire gene plus some sequences on either side of the transcribed region. The resulting molecules are examined in an electron microscope under conditions where regions of single- and double-stranded nucleic acids can be distinguished. The following structures are seen.

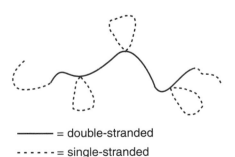

———— = double-stranded

- - - - - - = single-stranded

Of how many exons is this gene composed?

Answer Four.

13. The primary transcripts of eukaryotic class II genes are capped, polyadenylated and spliced. β-Thalassemia is a genetic disease caused by reduced expression of β-globin, the protein component of hemoglobin. One class of β-thalassemia results from a point mutation of the β-globin gene that changes the adenine nucleotide at an intron branch site to another nucleotide. Predict the consequences of this point mutation on globin pre-mRNA maturation and β-globin protein synthesis.

Answer

During pre-mRNA processing, spliceosomes cleave the 5′ end of an intron and covalently attach it to the 2′ hydroxyl of the adenosine residue of the branch site. A point mutation changing this adenosine to another nucleotide would block splicing of the intron, producing an altered mRNA that includes at least some intron sequences. Any attempts to translate such mRNAs would not yield functional β-globin, either because additional amino acids are encoded by the retained intron or because stop codons are encountered within the retained intron.

14. Transcriptional attentuation, like that observed for the *trp* operon of *E. coli*, is unlikely to occur in eukaryotes. Why?

Answer

Transcription attenuation control is made possible in prokaryotes by the mechanistic coupling between transcription and translation. Ribosomes translate a mRNA while it is being transcribed, with the first ribosome following closely behind RNA polymerase. In the case of *trp* operon attenuation, if insufficient aminoacylated-tRNA^{Trp} is available, the ribosomes pause at consecutive tryptophan codons in the *trp* leader region. The pausing of ribosomes at this site alters the intrastrand base-paired structure of *trp* leader sequences downstream of the pause site. The result is that a transcription-terminating stem-loop is not formed, and transcription proceeds past the attenuator site. If ribosomes do not pause at the critical tryptophan codons (due to sufficient levels of aminoacylated-tRNA^Trp), the transcription-terminating stem-loop forms and RNA polymerase terminates transcription at the attenuator site. In eukaryotes, transcription occurs in the nucleus, while translation occurs in the cytoplasm. Thus, transcription and translation are not coupled in eukaryotes. Attentuation control like that exemplified by the *trp* operon cannot occur in eukaryotes.

15. The direction that a ribosome translates mRNA was determined through the use of a cell-free protein synthesizing system which utilized an artificial mRNA consisting of 11 adenine bases followed by a 3′ terminal cytosine. The triplet codon AAA encodes the amino acid lysine (Lys), and the codon AAC encodes the amino acid asparagine (Asn). Diagram the resulting polypeptide indicating the carboxyl and amino termini.

Answer

The sequence of this synthetic message is 5′-AAA AAA AAA AAC-3′. This is equivalent to four triplet codons. Polypeptides are assembled from their amino termini to their carboxyl termini. Therefore, the sequence of the peptide corresponding to this synthetic oligonucleotide is NH_2-Lys-Lys-Lys-Asn-COOH.

16. How would a mutation deleting the Shine-Dalgarno region of an E. coli mRNA affect gene expression?

Answer

Ribosomes initiate translation near the 5′ end of a mRNA. Initiation of translation in *E. coli* is facilitated by complementary base pairing between a short region at the 3′ end of the 16S (small subunit) ribosomal RNA and the Shine-Dalgarno sequence, a region near the 5′ end of essentially all *E. coli* mRNAs. This base pairing is necessary for efficient translation of the mRNA. Without the Shine-Dalgarno region, the gene would be transcribed, but the mRNA would not be efficiently translated.

17. During translation elongation, an orderly sequence of events adds amino acids sequentially to the carboxyl terminus of a growing polypeptide chain. Interactions between tRNAs, mRNA, and the ribosome are instrumental to this process. Three different types of tRNA are utilized and/or produced during a cycle of translation elongation: aminocyl-tRNAs, peptidyl-tRNAs, and deacylated tRNAs. These tRNAs interact with three distinct tRNA binding sites on the ribosome: the A site, the E site, and the P site. Describe the associations of tRNAs with each of these tRNA binding sites during the following stages of translation elongation:

(a) Immediately before a "new" aminoacyl-tRNA is delivered to the ribosome.

(b) Immediately after a "new" aminoacyl-tRNA is delivered to the ribosome but before peptide bond formation.

(c) Immediately after peptide bond formation but before translocation.

(d) Immediately after translocation but before a "new" aminoacyl tRNA is delivered to the ribosome.

Answers

(a) A peptidyl-tRNA occupies the P site. The A site and E site are empty.

(b) A peptidyl-tRNA occupies the P site. An aminoacyl-tRNA occupies the A site. The E site is empty.

(c) A deacylated tRNA occupies the P site. A peptidyl-tRNA occupies the A site. The E site is empty.

(d) A peptidyl-tRNA occupies the P site. The A site is empty. The E site is occupied briefly by a deacylated tRNA as it exits the ribosome from the P site.

18. Compare and contrast the differing ways in which the 3′ ends of prokaryotic and eukaryotic mRNAs are produced.

Answer

The 3′ termini of prokaryotic mRNAs are formed by transcription termination. Two types of transcription termination occur in prokaryotes: Rho-independent and Rho-dependent termination. During Rho-independent termination, sequences present near the eventual 3′ end of a mRNA adopt a stem-loop structure during transcription due to intrastrand base pairing. This mRNA structure, together with a region of downstream oligo(U) residues disrupts the RNA polymerase/template complex, thereby releasing the nascent mRNA. In Rho-dependent termination, Rho protein binds to sites within the mRNA near its eventual 3′ end. Binding of Rho and subsequent ATP hydrolysis disrupt the RNA polymerase/template complex, thereby releasing the nascent transcript. Both Rho-independent and Rho-dependent termination in bacteria share the property that termination of transcription is an active process during which RNA polymerase is caused to cease transcription and to release the mRNA. In eukaryotes, however, the formation of mRNA 3′ termini does not involve cessation of transcription by RNA polymerase. Rather, RNA polymerase transcribes well beyond the site which will subsequently become the mature 3′ terminus. Specific sequences contained within the pre-mRNA near its eventual 3′ end cause the pre-mRNA to be cleaved at a nearby downstream site. A series of several dozen to several hundred adenosine residues (the poly(A) tract) is then added to the 3′ end liberated by cleavage of the pre-mRNA.

19. A region of mRNA sequence from a wild-type gene and that of four different mutant alleles is shown below. (The region shown is from the middle of a much larger mRNA.) What are the consequences of each mutation to the sequence of the encoded protein? You will need to consult a table of the genetic code (p. 233) to answer this question.

Wild-type mRNA: 5′-UCCCUGACCCAUGAAACCGCC-3′
Wild-type protein: Ser   Leu   Thr   His   Glu   Thr   Ala
    Mutation #1:   5′-UCCCUGAC**CU**AUGAAACCGCC-3′
    Mutation #2:   5′-UCCCUGACCCAU**U**AAACCGCC-3′
    Mutation #3:   5′-UCCCU**A**ACCCAUGAAACCGCC-3′
    Mutation #4:   5′-UCCCUGACCC**C**AUGAAACCGC-3′

Answer

Mutant #1:   Mutant #1 contains a single nucleotide substitution (C → U) that changes a CAU codon (histidine) to a UAU codon (tyrosine). The resulting protein has a single amino acid substitution relative to wild type. Such mutations are termed "missense" mutations.

Mutant #2:   Mutant #2 contains a single nucleotide substitution (G → U) that changes a GAA codon (glutamic acid) to a UAA codon (stop). The UAA stop codon causes translation to terminate prematurely, producing a protein that is truncated at the UAA mutation site. Such mutations are termed "nonsense" mutations.

Mutant #3: Mutant #3 contains a single nucleotide substitution (G → A) that changes a CUG codon (leucine) to a CUA codon (also leucine). The amino acid sequence of the resulting protein is identical to that of wild type.

Mutant #4: Mutant #4 contains a single nucleotide insertion (CCC → CCCC) that causes a change in reading frame of the mRNA. After shifting into a new reading frame, a UGA stop codon is encountered two codons after the insertion. The wild-type amino acid sequence Ser-Leu-Thr-His-Glu-Thr-Ala is changed to Ser-Leu-Thr-Pro-Stop, thus truncating the protein immediately downstream of the single base insertion. Such mutations are termed "frameshift" mutations.

20. During translation elongation, aminoacyl-tRNAs occupy the A site only if their anticodon:codon base pairing is correct. The rules for base pairing between the nucleotide located at the 5′ end of the anticodon and the 3′ end of the codon are special. A wider variety of base pairing is allowed at this "wobble" position than elsewhere in the codon or, more generally, elsewhere in double-stranded nucleic acids. Transfer RNAs, furthermore, contain numerous modified bases that are made enzymaticlaly after a tRNA is synthesized. One such modified base, inosine, is often found at the 5′ (wobble) position of an anticodon. The following table describes allowable base pairing at the wobble position.

Base at the 5′ end of an anticodon	Bases recognized at the 3′ end of a codon
U	A or G
C	G only
A	U only
G	C or U
I	U, C, or A

For each of the following anticodons, list all of the codons with which the tRNA will base pair during translation. What amino acids are encoded by each codon? You will need to consult a table of the genetic code (p. 233) to answer this question.

(a) 5′-UAG -3′

(b) 5′-CCA-3′

(c) 5′-GAU-3′

(d) 5′-ICG-3′

(e) 5′-ACC-3′

Answer

(a) 5′-CUA-3′ and 5′-CUG-3′ (both encode leucine)
(b) 5′-UGG-3′ (tryptophan)
(c) 5′-AUC-3′ and 5′-AUU-3′ (both encode isoleucine)
(d) 5′-CGU-3′, 5′-CGC-3′, and 5′-CGA-3′ (all three encode arginine)
(e) 5′-GGU-3′ (glycine)

Note that whenever wobble causes multiple codons to be read, the same amino acid is inserted in each case. This is an important property of wobble.

21. The sequence of a bacterial gene encoding a small protein is shown below.

5′-CGATGCATGGACGGGGCTTAGCCTAAGG-3′
3′-GCTACGTACCTGCCCCGAATCGGATTCC-5′

(a) If transcription proceeds from left to right and begins at the first nucleotide shown above, what is the sequence of the resulting mRNA? Indicate the 5′ and 3′ termini.
(b) Assume that translation initiates at the first methionine codon of the mRNA. What is the sequence of the protein encoded by this gene? Indicate the amino and carboxyl termini. You will need to consult a table of the genetic code (p. 233) to answer this and subsequent questions.
(c) If the G:C base pair at position 9 of the gene (number 1 is at the left) is mutated to an A:T base pair, what will the sequence of the resulting protein be?
(d) If the G:C base pair at position 14 of this gene were deleted (a -1 frameshift mutation), what will the sequence of the resulting protein be?

Answers

(a) 5′-CGAUGCAUGGACGGGGCUUAGCCUAAGG-3′
(b) NH_2-Met-His-Gly-Arg-Gly-Leu-Ala-COOH
(c) NH_2-Met-His-Arg-Arg-Gly-Leu-Ala-COOH
(d) NH_2-Met-His-Gly-Arg-Ala-COOH

22. Fill in the blanks in the following table. You will need to consult a table of the genetic code (p. 233) to answer the question. Do not consider wobble base pairing in your answers.

DNA double helix	5'-				A				G				G	-3'
	3'-			G						T	T			-5'
mRNA	5'-							U	C					-3'
tRNA anticodon	3'-				G			C						-5'
amino acids	NH$_2$	MET			GLU							-COOH

Answer

DNA double helix	5'-	A	T	G	C	A	C	G	A	G	T	G	C	A	A	G	-3'
	3'-	T	A	C	G	T	G	C	T	C	A	C	G	T	T	C	-5'
mRNA	5'-	A	U	G	C	A	C	G	A	G	U	G	C	A	A	G	-3'
tRNA anticodon	3'-	U	A	C	G	U	G	C	U	C	A	C	G	U	U	C	-5'
amino acids	NH$_2$	MET		HIS		GLU		CYS		LYS		-COOH

23. The following RNA sequence is taken from the middle of a much larger messenger RNA. The entire region shown below encodes part of a protein.

5′-GACGAGCACUAACGCAUAGGCCAGUA-3′

What is the sequence of amino acids encoded by this portion of the mRNA? You will need to consult a table of the genetic code (p. 233) to answer this question.

Answer

(NH$_2$ terminus)-Arg-Ala-Leu-Thr-His-Arg-Pro-Val-(COOH terminus)

24. At what point during translation is the ribosomal P site occupied by an aminoacyl-tRNA (as opposed to a peptidyl-tRNA or a deacylated tRNA)?

Answer

Only during initiation. The first aminoacyl-tRNA that interacts with the ribosome is methionyl-tRNA^{Met} (eukaryotes) or N-formyl-methionyl-tRNA^{Met} (prokaryotes). During translation initiation, aminoacyl-tRNA^{Met} associates with the portion of the P site contained on the ribosomal small subunit. Aminoacyl-tRNA^{Met} assembles with mRNA, translation initiation factors, the ribosomal small subunit, and GTP to form a translation initiation complex. After a ribosomal large subunit joins the complex, aminoacyl-tRNA^{Met} is bound at the now-complete P site. After an appropriate aminoacyl-tRNA is delivered to the A site, the first peptide bond is formed. Thereafter, and during all of translation elongation, tRNAs bound to the P site are either peptidyl-tRNAs (present after translocation but before peptide bond formation) or deacylated tRNAs (present after peptide bond formation but before translocation).

25. Ferritin is an iron-binding protein present in red blood cells (mature red blood cells lack nuclei). When iron is abundant, the half-life of ferritin mRNA is 48 hours; if iron is scarce, the half-life of ferritin mRNA is 12 hours. The half-life of a mRNA is the time required for the cell to degrade half of the mRNA present in the cell at any point in time. How does the abundance of iron affect the accumulation of ferritin protein in red blood cells?

Answer

The amount of protein produced is proportional to the steady-state level of its mRNA. When iron is abundant, the ferritin mRNA is long-lived and may be translated many times. When iron is scarce, the ferritin mRNA is degraded more rapidly and, therefore, the red cells produce less ferritin protein.

Universal Genetic Code

Second position of mRNA codon

		U	C	A	G	
	U	U U U Phe U U C U U A Leu U U G	U C U U C C Ser U C A U C G	U A U Tyr U A C U A A Stop U A G Stop	U G U Cys U G C U G A Stop U G G Trp	U C A G
	C	C U U C U C Leu C U A C U G	C C U C C C Pro C C A C C G	C A U His C A C C A A Gln C A G	C G U C G C Arg C G A C G G	U C A G
	A	A U U A U C Ile A U A A U G Met	A C U A C C Thr A C A A C G	A A U Asn A A C A A A Lys A A G	A G U Ser A G C A G A Arg A G G	U C A G
	G	G U U G U C Val G U A G U G	G C U G C C Ala G C A G C G	G A U Asp G A C G A A Glu G A G	G G U G G C Gly G G A G G G	U C A G

First position (5' end) of mRNA codon

Third position (3' end) of mRNA codon

☐ Initiation codon   ▦ Termination codons

Control of
Gene Expression

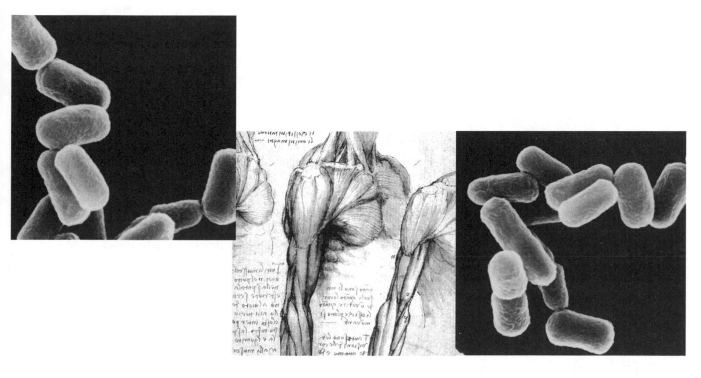

Summary

Regulation of gene expression is an important part of how cells respond to their environment. The identity of a cell and the functions it performs are dictated in large part by the collection of proteins it expresses. Patterns of gene expression change as organisms grow and differentiate. Genetic tools have been used to unravel the control mechanisms. In both prokaryotes and eukaryotes, a diversity of regulatory proteins control gene expression. Such proteins modulate the rate at which individual genes are transcribed, the manner in which their mRNAs are processed, the efficiencies with which the mRNAs are translated, and the rates at which the mRNAs or proteins are degraded within the cell.

Control of gene expression in bacteria most often occurs by controlling the rate at which *RNA polymerase* binds to a promoter and initiates transcription. Various factors influence polymerase binding and the initiation of transcription. Small proteins called σ factors associate with the core RNA polymerase and direct the polymerase complex to bind to specific promoters. Other proteins may bind to specific DNA sequences near the promoters, either to accelerate transcription or to inhibit the movement of the polymerase from the promoter. For example, the *lac* **repressor**, *lac i* for *inhibitor*, binds to *lac o*, the **operator**, of the *lac* operon. *Lac o* is immediately adjacent to *lac p*, the *lac* promoter. If the repressor is bound to the operator, RNA polymerase is prevented from transcribing the operon. If lactose is present in the bacterial medium, a by-product of lactose binds to the repressor, preventing it from binding to the operator, and RNA polymerase proceeds to transcribe the *lac* operon. Repressor-mediated regulation of the *lac* operon is an example of **negative control**: transcription is inhibited, until **induced** by lactose.

The rate of transcription of the *lac* operon may vary up to 1000-fold. When both glucose and lactose are present as fuels, bacteria preferentially use glucose, and transcription of the *lac* operon is low. When glucose is absent, the intracellular concentration of *cyclic adenosine monophosphate, cAMP,* rises. cAMP binds to the **catabolite activator protein,** *CAP.* The cAMP-CAP complex binds upstream of the *lac* promoter and stimulates RNA polymerase binding, thereby maximizing *lac* operon transcription. This is an example of **positive control** of gene expression.

Regulation of expression of the *trp* operon involves both negative and positive control. When tryptophan is plentiful, *trpR,* the repressor, binds to *trpO*, the operator, and prevents *trp* operon transcription. Under such conditions, free tryptophan (the **corepressor**) binds directly to the repressor protein, thereby inducing a conformational change that allows the repressor to bind the operator. When tryptophan is limiting, the repressor is bound to neither tryptophan nor the operator. This is an example of negative control. Expression of the *trp* operon is under positive control by transcriptional **attenuation.** This regulatory mechanism relies on the coupling of transcription and translation in prokaryotic cells. At the 5' end of the *trp* operon, there is a short open reading frame called the **leader sequence.** Intrastrand base pairing of the RNA in the leader sequence forms several stem-loop structures. One stem-loop structure, if formed, signals transcription termination. The *trp* leader sequence is translated as it is being transcribed. As a ribosome translates the leader sequence, it encounters several tryptophan codons. If there is sufficient aminoacyl-tRNA^{Trp}, the ribosome proceeds rapidly through the leader, and the transcription-terminating stem-loop forms, dislodging the RNA polymerase and prematurely terminating *trp* operon transcription. However, if aminoacyl-tRNA^{Trp} is limiting, the ribosome stalls at the tryptophan codons, the transcription-terminating stem-loop does not form, and RNA polymerase proceeds to transcribe the operon.

Gene expression in eukaryotes is regulated at several stages, including transcription, translation, mRNA processing, and mRNA degradation. During transcriptional control, large complexes of transcription-associated factors, or TAFs, form at or near the TATA box of eukaryotic promoters. General transcription factors join the complex and recruit type II RNA polymerase, which is also a large protein complex.

Most eukaryotic genes are under positive control. Transcription factors associate with each other and with the basal transcription machinery through protein-protein interactions. **Enhancers** are *cis*-acting regulatory elements located within DNA to which transcription-activating factors bind. Activator proteins have characteristic DNA-binding domains that recognize specific enhancers. Enhancers are active in either orientation relative to the genes they regulate and may be located at considerable distance from those genes. Developmental and tissue-specific regulation is controlled, in part, by the interaction of signal molecules (i.e., steroid hormones) with cognate-binding proteins. Such complexes bind to specific enhancers, (i.e., hormone response elements) to stimulate transcription of specific genes. Localized changes in nucleosome and chromatin structure are associated with active gene transcription.

Control of gene expression is often regulated post-transcriptionally in eukaryotes. The manner in which pre-mRNAs are spliced is frequently regulated. For example, alternative splicing of a single Drosophila tropomyosin gene produces the distinct forms of tropomyosin found in either embryos or adults. The rate at which mRNAs are translated can also be regulated. For example, translation of ferritin mRNA (an iron storage protein) increases under conditions of elevated intracellular iron. Increased translation of ferritin mRNA increases iron storage capacity and decreases iron toxicity within cells. The rate at which mRNAs are degraded can be regulated. For example, the transferrin receptor mRNA (an iron transport protein) is more rapidly degraded under conditions of elevated intracellular iron. Finally, the rate at which proteins are degraded can be regulated. For example, cyclin (a cell cycle regulatory protein) is rapidly degraded at specific times during the cell cycle.

Self-Testing Questions

1. Indicate whether the following statements are true or false.
    (a) A mutation of CAP that reduces its affinity for cyclic AMP causes reduced expression of *lac* operon enzymes.
    (b) *E. coli* grown in glucose have high concentrations of cyclic AMP. Those grown in glycerol (or other non-fermentable carbon sources) have low concentrations of cyclic AMP.
    (c) In the presence of excess tryptophan, the *trp* attenuator region is transcribed more often than the structural genes encoding tryptophan biosynthetic enzymes.
    (d) RNA polymerase III is responsible for transcribing rRNA genes in eukaryotes.
    (e) Steroid hormone receptor proteins can be located in either the cytoplasm or the nucleus of a cell.
    (f) Eukaryotic enhancers can be located either upstream or downstream of the transcriptional start sites of the genes they regulate.
    (g) Transcription factor IID (TFIID), a component of the eukaryotic basal transcription machinery, binds to the promoter sequence 5'-CCAAT-3' ("CAAT box").

Answer

T   **(a)** A mutation of CAP that reduces its affinity for cyclic AMP causes reduced expression of *lac* operon enzymes.

F   **(b)** *E. coli* grown in glucose have high concentrations of cyclic AMP. Those grown in glycerol (or other non-fermentable carbon sources) have low concentrations of cyclic AMP.

T   **(c)** In the presence of excess tryptophan, the *trp* attenuator region is transcribed more often than the structural genes encoding tryptophan biosynthetic enzymes.

F   **(d)** RNA polymerase I is responsible for transcribing rRNA genes in eukaryotes.

T   **(e)** Steroid hormone receptor proteins can be located in either the cytoplasm or the nucleus of a cell.

T   **(f)** Eukaryotic enhancers can be located either upstream or downstream of the transcriptional start sites of the genes they regulate.

F   **(g)** Transcription factor IID (TFIID), a component of the eukaryotic basal transcription machinery, binds to the promoter sequence 5′-CCAAT-3′ ("CAAT box").

2. Fill in the blank with the defined word or phrase.
  (a) The DNA site to which RNA polymerase binds specifically and within which transcription is initiated
  (b) The DNA site within the *lac* operon to which *lac* repressor binds
  (c) A small effector molecule that stimulates the transcriptional activity of a bacterial operon by binding to and inactivating a repressor protein
  (d) A regulatory sequence of DNA in eukaryotes that controls the abundance, tissue specificity, and temporal pattern of transcription of an associated gene. Such sequences exert their effects through their sequence-specific binding of regulatory proteins that stimulate activity of RNA polymerase bound at nearby promoters.
  (e) The generation of different mature mRNAs from a single eukaryotic gene by splicing together different sets of exons derived from the same primary transcript

Answer
(a) promoter
(b) operator
(c) inducer
(d) enhancer
(e) alternative splicing

3. Complete the following table by entering the quantity of β-galactosidase expressed by the indicated genotypes. An entire *lac* operon (including *lacI*) is present on both the chromosome and any F′ elements. If a *lac* gene or regulatory site is not indicated in a genotype, it is unaffected by mutation. IPTG is a potent inducer of *lac* operon expression. Enter either < 0.01, 2, or 1,000 in the table.

	β-GALACTOSIDASE ENZYME (UNITS)	
Genotype	**–IPTG**	**+IPTG**
wild type (*lac⁺*)	2	1000
lacZ⁻	<0.01	<0.01
lacI⁻	_____	_____
lacI⁻/F′ lac⁺	_____	_____
lacP⁻	_____	_____
lacP⁻/F′ lacZ⁻	_____	_____
lacIˢ	_____	_____
lacIˢ/F′ lacI⁺	_____	_____
lacOᶜ	_____	_____
lacOᶜ/F′ lacIˢ	_____	_____

➤ **Solution**

Answer:

	β-GALACTOSIDASE ENZYME (UNITS)	
Genotype	**–IPTG**	**+IPTG**
lacI⁻	1,000	1,000
lacI⁻/F′ lac⁺	2	1,000
lacP⁻	<0.01	<0.01
lacP⁻/F′ lacZ⁻	<0.01	<0.01
lacIˢ	2	2
lacIˢ/F′ lacI⁺	2	2
lacOᶜ	1,000	1,000
lacOᶜ/F′ lacIˢ	1,000	1,000

lacI⁻   *LacI* encodes the *lac* repressor. In the absence of the repressor, the *lac* operon is transcribed at maximal levels in both the presence and absence of the inducer.

lacI⁻/F′ lac⁺   Because *lacI⁻* mutations eliminate the *lac* repressor, they are recessive. Expression of the *lac* operon is normal in a *lacI⁻/F′ lac⁺* merodiploid.

lacP⁻   *LacP* is the *lac* operon promoter, the site to which RNA polymerase binds during initiation of transcription. Without a promoter, the *lac* operon is unexpressed under all conditions.

$lacP^-/F'lacZ^-$	The only copy of $lacZ^+$ in this strain is located on the chromosome. The chromosomal *lac* operon is $lacP^-$. LacP, the *lac* operon promoter, is a regulatory site in the DNA. Its effects, therefore, are *cis*-dominant. LacZ is unexpressed under all conditions in this strain.
$lacI^S$	$LacI^S$ encodes a *lac* "super repressor" that is insensitive to the presence of the inducer. Such mutant repressors bind to the *lac* operator and repress operon expression even in the presence of the inducer.
$lacI^S/F'lacI^+$	Because the $lacI^S$ repressor is insensitive to inducer, its effects are *trans*-dominant. Expression of the *lac* operon is low and constitutive in this merodiploid strain.
$lacO^C$	LacO is the *lac* operator, the site to which *lac* repressor binds. Operator constitutive mutations ($lacO^C$) do not bind the repressor under any conditions. Thus, the *lac* operon is expressed at maximal levels under all conditions in $lacO^C$ mutants.
$lacO^C/F'lacI^S$	Because the *lac* operator is a regulatory site in DNA, its effects are *cis*-dominant. Even though $LacI^S$ encodes a repressor that is insensitive to the inducer, the $lacI^S$ repressor cannot bind the $lacO^C$ operator. Thus, the *lac* operon is expressed at maximal levels under all conditions.

4. *Lac-6* is an allele of the *lac* operon. *Lac-6* was transduced onto an F'*lac* element by generalized transduction, and the resulting F'*lac-6* was introduced by conjugation into strains containing known chromosomal *lac* mutations. The uninduced and induced quantities of β-galactosidase expressed in the F'-containing strains are shown in the table below.

	β-GALACTOSIDASE ENZYME LEVEL	
Genotype	Uninduced	Induced
lac^+ (wild type)	2	1,000
$lac^+/F'lac-6$	2	1,000
$lacZ^-/F'lac-6$	<1	<1
$lacI^-/F'lac-6$	2	1,000
$lacO^c/F'lac-6$	1,000	1,000
$lacI^S/F'lac-6$	2	2

What type of *lac* allele is *lac-6*?

Answer Either $lacZ^-$ or $lacP^-$

5. Complete the following table to indicate whether β-galactosidase (encoded by *lacZ*) and *lac* permease (encoded by *lacY*) is expressed and, if expressed, whether its expression is constitutive or inducible. Enter "A" (absent), "C" (constitutive, either high or low), or "I" (inducible) in the table.

Genotype	Expression of *lacZ*	Expression of *lacY*
lacI⁻ lacP⁺O^C Z⁺Y⁻/F'lacI⁻ lacP⁺O⁺Z⁻Y⁺	——	——
lacI⁻ lacP⁺O⁺Z⁺Y⁻/F'lacI⁺ lacP⁺O^C Z⁻Y⁺	——	——
lacI⁺ lacP⁺O^C Z⁺Y⁻/F'lacI⁺ lacP⁺O⁺Z⁻Y⁻	——	——
lacI⁺ lacP⁺O^C Z⁺Y⁻/F'lacI⁻ lacP⁺O⁺Z⁻Y⁺	——	——
lacI^S lacP⁻O⁺Z⁻Y⁻/F'lacI⁻ lacP⁺O⁺Z⁺Y⁻	——	——
lacI⁺ lacP⁻O^C Z⁺Y⁻/F'lacI⁻ lacP⁺O^C Z⁻Y⁺	——	——

Answer

Genotype	Expression of *lacZ*	Expression of *lacY*
lacP⁺O^C Z⁺Y⁻/F'lacI⁻ lacP⁺O⁺Z⁻Y⁺	C	C
lacI⁻ lacP⁺O⁺Z⁺Y⁻/F'lacI⁺ lacP⁺O^C Z⁻Y⁺	I	C
lacI⁺ lacP⁺O^C Z⁺Y⁻/F'lacI⁺ lacP⁺O⁺Z⁻Y⁻	C	A
lacI⁺ lacP⁺O^C Z⁺Y⁻/F'lacI⁻ lacP⁺O⁺Z⁻Y⁺	C	I
lacI^S lacP⁻O⁺Z⁻Y⁻/F'lacI⁻ lacP⁺O⁺Z⁺Y⁻	C	A
lacI⁺ lacP⁻O^C Z⁺Y⁻/F'lacI⁻ lacP⁺O^C Z⁻Y⁺	A	C

6. As part of your studies concerning arginine catabolism in bacteria, you have isolated four mutants (*arg-1*, *-2*, *-3*, and *-4*) affecting expression of arginase, an enzyme required for breakdown of arginine. The quantity of arginase expressed by wild type and by the mutants is shown in the table below

Min = minimal medium; Arg = arginine added to the medium

	Min	Min+Arg
wild type	10	500
arg-1	50	50
arg-2	10	100
arg-3	500	500
arg-4	500	10

In which mutant(s) is arginase expressed constitutively?

➤ ***Solution***

Answer: arg-1 and arg-3

Constitutive expression means *unchanging* expression. Genes are expressed constitutively when the level of expression is the same under all conditions tested. Constitutive expression does not imply anything about the level of expression. Genes can be expressed constitutively at high level, low level, or anything in between. Mutations *arg-1* and *arg-3* are expressed constitutively.

7. Mutations *lac-1*, *lac-2*, and *lac-3* (abbreviated *1*, *2*, and *3* in the table) alter regulation of the *lac* operon. One of these mutations affects the *lac* repressor, one the *lac* promoter, and one the *lac* operator. The quantities of β-galactosidase expressed by various haploid and merodiploid strains are shown in the table below.

	β-GALACTOSIDASE ACTIVITY	
Genotype	Uninduced	Induced
$1^+2^+3^+$	Low	High
$1^-2^+3^+$	High	High
$1^+2^-3^+$	High	High
$1^+2^+3^-$	Low	Low
$1^-2^+3^+/F' 1^+2^+3^-$	High	High
$1^+2^-3^+/F' 1^-2^+3^-$	Low	High

Which of these mutations affects the *lac* repressor? The promoter? The operator?

Answer

lac-1, *lac-2*, and *lac-3* affect the operator, repressor, and promoter, respectively.

8. Catabolite repression in *E. coli* is mediated by two regulatory molecules: the Catabolite Activator Protein (CAP) and cyclic AMP (cAMP). CAP is encoded by the *crp* gene; cAMP is synthesized by adenyl cyclase, encoded by the *cya* gene. lac^{UV5} is a *lac* promoter mutation that renders *lac* operon expression independent of CAP-cAMP. Complete the following table by entering the quantity of β-galactosidase expressed by the indicated genotypes. An entire *lac* operon is present on both the chromosome and any F' elements. If a *lac* gene or regulatory site is not indicated in a genotype, it is unaffected by mutation. IPTG is an inducer of *lac* operon expression. Enter either 2, 20, 350, or 1,000 in the table.

	QUANTITY OF β-GALACTOSIDASE			
	GLYCEROL-GROWN CELLS		GLUCOSE-GROWN CELLS	
Genotype	−IPTG	+IPTG	−IPTG	+IPTG
wild type (*lac⁺*)	2	1,000	2	350
$crp^-\ cya^-$	2	20	2	20
lac^{UV5}	2	1,000	2	1,000
$crp^-\ cya^-\ lac^{UV5}$	____	____	____	____
$lac^{UV5}/F'\,lacZ^-$	____	____	____	____

➤ *Solution*

Answer:

| | QUANTITY OF β-GALACTOSIDASE | | | |
| | GLYCEROL-GROWN CELLS | | GLUCOSE-GROWN CELLS | |
Genotype	−IPTG	+IPTG	−IPTG	+IPTG
$crp^-\ cya^-\ lac^{UV5}$	2	1,000	2	1,000
$lac^{UV5}/F'lacZ^-$	2	1,000	2	1,000

"Catabolite repression" is a regulatory system of bacteria that causes cells to utilize glucose in lieu of many other carbon sources. When glucose is available in the medium, intracellular levels of cyclic AMP (cAMP) are low. When glucose is absent in the medium, cAMP levels are high. Maximal levels of cAMP occur when nonfermentable carbon sources, such as glycerol, are the only ones available.

The effects of catabolite repression are exerted by the CAP-cAMP complex, a positive regulatory factor required for expression of many genes, including the *lac* operon. (The term catabolite *repression* is somewhat of a misnomer. The presence of glucose in the growth medium *represses* gene expression, but the mechanism involves *activation* of gene expression in the absence of glucose.) Catabolite repression is evident in two different places in the table. First, wild-type cells induced on glucose express only about one third the level of β-galactosidase as those induced on glycerol (line 1). Second, $crp^-\ cya^-$ double mutants, in which a CAP-cAMP complex is not formed, express low levels of β-galactosidase (line 2).

The mutation lac^{UV5} renders *lac* operon expression insensitive to catabolite repression. Expression of lac^{UV5} is induced to maximal levels even when cells are grown on glucose (line 3). Thus, expression of lac^{UV5} no longer requires the CAP-cAMP complex. If expression no longer requires CAP-cAMP, then crp^- and cya^- mutations should not influence expression of lac^{UV5}. Thus, expression of β-galactosidase in $crp^-\ cya^-\ lac^{UV5}$ triple mutants (line 4) is identical to that of lac^{UV5} single mutants. In a strain of genotype $lac^{UV5}/F'lacZ^-$ (line 5), the only $lacZ^+$ allele is on the chromosome, linked to the lac^{UV5} mutation. Because lac^{UV5} affects the *lac* promoter, it is *cis*-dominant. Thus, expression of β-galactosidase in the merodiploid strain is identical to that of lac^{UV5}.

9. In the bacterium *Coca coli*, expression of the tryptophan operon is under repression control, similar to that of *E. coli*. You have isolated $trpR^a$, an allele of the *C. coli* tryptophan repressor gene. $TrpR^a$ mutants are tryptophan auxotrophs because they fail to express the *trp* operon even when cells are grown in the absence of tryptophan.

(a) What is the most likely molecular defect of $trpR^a$ mutants?

(b) Predict the phenotype of $trpR^a/F'trpR^+$ merodiploids.

Answers

(a) $TrpR^a$ alters the repressor such that it binds to the trp operator and represses expression even when the corepressor (tryptophan) is absent.

(b) $TrpR^a/F'trpR^+$ merodiploids will be unable to express the trp operon even in the absence of tryptophan. They will be tryptophan auxotrophs.

10. Transcription of the trp operon is regulated by two independent systems. Regulatory components include: $trpR$, the repressor; $trpO$, the operator; $trpP$, the promoter; $trpL$, the leader; and $trpa$, the attenuator. $TrpR$ is unlinked to the trp operon. $TrpE$ encodes a subunit of anthranilate synthetase, an enzyme required for tryptophan biosynthesis. Complete the following table by entering the quantity of anthranilate synthetase expressed by the indicated genotypes. An entire trp operon (but not including $trpR$) is present on both the chromosome and any $F'trp$ elements. If a trp gene or regulatory site is not indicated in a genotype, it is unaffected by mutation. Enter either <0.01, 1, 10, 70, or 700 in the table.

Genotype	Relative Amount of Anthranilate Synthetase
wild type (trp^+)	1
$trpP^-$	<0.01
$trpa^-$	10
$trpR^-$	70
$trpO^c$	70
$trpR^-\ trpa^-$	700
$trpO^c a^-$	____
$trpR^-\ trpa^-/F'trpR^+$	____
$trpa^-E^+/F'trpa^+E^-$	____

Answer

Genotype	Relative Amount of Anthranilate Synthetase
$trpO^c a^-$	700
$trpR^-\ trpa^-/F'trpR^+$	10
$trpa^-E^+/F'trpa^+E^-$	10

11. Which of the following regulatory mutations are dominant to the corresponding wild-type allele? Check all that apply.

_____ *lacO^C*, an operator constitutive mutation of the *lac* operon
_____ *crp*, a mutation of the Catabolite Activator Protein (CAP) that prevents the allosteric transition of CAP which normally occurs upon binding of cAMP to CAP
_____ *lacI^S*, a mutation of the *lac* repressor making it insensitive to inducer
_____ *trpa^−*, a mutation eliminating the *trp* operon attenuator
_____ *araI^−*, a mutation that destroys function of the arabinose operon initiator
_____ *lacP^−*, a mutation affecting the *lac* promoter that reduces its affinity for the CAP-cAMP complex but which does not affect binding of RNA polymerase to the promoter

Answer

__X__ *lacO^C*, an operator constitutive mutation of the *lac* operon
_____ *crp*, a mutation of the Catabolite Activator Protein (CAP) that prevents the allosteric transition of CAP which normally occurs upon binding of cAMP to CAP
_____ *lacI^S*, a mutation of the *lac* repressor making it insensitive to inducer
__X__ *trpa^−*, a mutation eliminating the *trp* operon attenuator
__X__ *araI^−*, a mutation that destroys function of the arabinose operon initiator
__X__ *lacP^−*, a mutation affecting the *lac* promoter that reduces its affinity for the CAP-cAMP complex but which does not affect binding of RNA polymerase to the promoter

12. Genes *enzA, enzB, enzC,* and *enzD* encode enzymes involved either in the synthesis or breakdown of compound "Z." The relative quantities of *enzA, B, C,* and *D* mRNA, protein, and enzymatic activity are shown in the table below.

	RELATIVE QUANTITY OF					
	mRNA		PROTEIN		ENZYMATIC ACTIVITY	
	Min	Min+Z	Min	Min+Z	Min	Min+Z
enzA	10	1	10	1	10	1
enzB	1	1	5	1	20	1
enzC	1	20	1	50	1	50
enzD	10	1	10	1	10	1

Min indicates cells grown in minimal medium. Min+Z indicates cells grown in minimal medium to which compound Z is added. The numbers shown are proportional to the measured quantity of mRNA (molecules per cell), protein (molecules per cell), or enzymatic activity (units of enzyme activity per cell measured *in vivo* in growing cells). Assume that compound Z does not affect turnover (degradation) of either mRNA or protein.

(a) Compound Z (or a metabolite of compound Z) is an inducer involved in transcriptional regulation of which gene(s)?

(b) Compound Z (or a metabolite of compound Z) is a corepressor involved in transcriptional regulation of which gene(s)?

(c) Which enzyme(s) are subject to feedback inhibition by compound Z?

(d) Expression of which gene(s) are subject to translational regulation by compound Z?

➤ *Solutions*

(a) *Answer: enzC*

Inducers are low molecular weight compounds that increase transcription of a gene by interfering with a repressor protein. To evaluate whether compound Z is an inducer of gene expression, we should compare the quantity of mRNA expressed in the presence and absence of compound Z. The quantity of mRNA reflects transcriptional activity of a gene more accurately than the quantity of protein or enzymatic activity because it is uncomplicated by effects that compound Z might have on translational efficiency or enzymatic activity. For which genes is mRNA abundance increased in the presence of compound Z? mRNA levels of *enzA*, *enzB*, and *enzD* decrease or remain unchanged in the presence of Z, but abundance of *enzC* mRNA increases. The increase is evident not only in the levels of *enzC* mRNA, but also in the levels of protein and enzymatic activity. Thus, compound Z induces *enzC* gene expression.

(b) *Answer: enzA and enzD*

Corepressors are low molecular weight compounds that decrease transcription of a gene by activating a repressor protein. As discussed in part **(a)**, the quantity of mRNA most accurately reflects transcriptional activity of a gene. For which genes is mRNA abundance decreased in the presence of compound Z? mRNA levels of *enzB* and *enzC* increase or remain unchanged, but abundance of *enzA* and *enzD* mRNA decreases. Thus, compound Z is a corepressor for expression of *enzA* and *enzD*.

(c) *Answer: enzB*

Feedback inhibitors are low molecular weight compounds that bind to an enzyme and directly inhibit its biochemical activity. Feedback inhibitors are often the endproduct of a biosynthetic pathway, and the targets of inhibition are usually an enzyme involved in its synthesis. The enzymatic activities of enzymes A, B, C, and D all decrease when cells are grown in the presence of compound Z, but it is important to recognize which of these decreases are due to changes in gene expression and which are due to differences in activity of the enzyme. For example, activity of enzyme A is 10-fold lower when cells are grown in the presence of compound Z. However, the quantity of *enzA* protein is also 10-fold lower. The inherent biochemical activity of the enzyme (activity per molecule of enzyme) is the same when cells are grown in the presence and absence of compound Z. Similarly, activity of enzyme D directly reflects the quantity of *enzD* protein. Activity of enzyme B, however, is different. Cells grown in the presence of compound Z contain 5-fold less *enzB protein*, but they contain 20-fold less *enzymatic activity*. Thus, enzyme B is feedback inhibited by compound Z.

(d) *Answer:* enzB *and* enzC

"Translational regulation" refers to modulation of the efficiency with which mRNA of an individual gene is translated. If all mRNAs were translated with equal efficiency, the amount of a specific protein would be directly proportional to the amount of its mRNA. mRNAs, however, are not translated with equal efficiency. Some are translated efficiently (producing many molecules of protein per molecule of mRNA), and some are translated inefficiently. The efficiency with which specific mRNAs are translated can be influenced by small molecules, such as compound Z. The amounts of *enzA, B, C,* and *D* protein all change when cells are grown in the presence of compound Z, but it is important to recognize which of these decreases are due to changes in mRNA abundance and which are due to changes in translational efficiency. For example, the quantity of enzyme A protein is 10-fold lower when cells are grown in the presence of compound Z. The quantity of *enzA* mRNA, however, is also 10-fold lower. The amount of enzyme A protein directly reflects the amount of *enzA* mRNA under both conditions. Thus, *enzA* mRNA is translated with equal efficiency in cells grown in the presence and absence of compound Z. Similarly, the quantity of enzyme D protein directly reflects the quantity of *enzD* mRNA. Enzymes B and C, however, are different. *EnzB* mRNA levels are unaffected by the presence of compound Z, but enzyme B protein is 5-fold lower in cells grown in the presence of Z. Thus, compound Z reduces the efficiency with which *enzB* mRNA is translated. Similarly, compound Z increases the efficiency with which *enzC* mRNA is translated.

---

13. *BelA* encodes an enzyme required to degrade the hypothetical sugar "bellicose." You have isolated four regulatory mutations, *bel-1, -2, -3,* and *-4,* that alter expression of *belA* but do not affect the coding region of the *belA* enzyme itself. All three mutations are tightly linked to *belA.* The relative quantities of *belA* enzyme expressed by various strains are shown in the table below.

	CARBON SOURCE	
Genotype	Glycerol	Bellicose
Wild Type	1	500
bel-1	500	500
bel-2	1	1
bel-3	500	500
bel-4	1	1
bel-1/F′belA⁻	500	500
bel-2/F′belA⁻	1	500
bel-3/F′belA⁻	1	500
bel-4/F′belA⁻	1	1

Is expression of *belA* regulated positively, negatively, or both positively and negatively by *trans*-acting regulatory factors?

➤ **Solution**

Answer: both positively and negatively

Positive regulation occurs when *trans*-acting regulatory proteins activate or increase expression of a gene. Positive regulatory factors usually bind to sites adjacent to a regulated gene and promote initiation of nearby transcription by RNA polymerase bound at the promoter. Elimination of a *trans*-acting positive factor by mutation causes the target gene to be unexpressed (or expressed at lowered levels). Such mutations are almost always recessive. This distinguishes them from other types of regulatory mutations that cause lowered expression, such as promoter mutations or mutations affecting eukaryotic enhancers. Promoter and enhancer mutations are *cis*-dominant, because they affect regulatory *sites*, not regulatory *proteins*. Are there any recessive mutations in the table that reduce *belA* expression? Yes, expression of *belA* is uninducible in *bel-2* mutants (line 3), and *bel-2* is recessive (line 7).

Negative regulation occurs when *trans*-acting regulatory proteins repress or inhibit expression of a gene. Negative regulatory factors usually bind to sites adjacent to a regulated gene and inhibit initiation of nearby transcription by RNA polymerase. Elimination of a *trans*-acting negative factor by mutation causes the target gene to be expressed at high levels or under inappropriate conditions. Such mutations are almost always recessive. This distinguishes them from other types of regulatory mutations that cause elevated expression, such as operator-constitutive mutations or mutations affecting a eukaryotic silencer. Operator-constitutive and silencer mutations are *cis*-dominant because they affect regulatory *sites*, not regulatory *proteins*. Are there any recessive mutations in the table that increase *belA* expression? Yes, expression of *belA* is constitutive at high levels in *bel-3* mutants (line 4), and *bel-3* is recessive (line 8).

14. Initiation of transcription in eukaryotes is regulated by a diversity of DNA-binding transcription factors that either activate or inhibit initiation of transcription. The DNA sequences to which positive or negative transcription factors bind act in *cis*, and can be located at considerable distance (either upstream or downstream) from the transcription start site. Negative transcription factors (those that inhibit transcription) bind to DNA sequence elements often called "silencers." What is the name of the DNA sequence elements to which positive transcription factors (those that activate transcription) bind?

Answer
Upstream Activating Sequences (yeast) or enhancers (most other eukaryotes)

15. The hormone insulin regulates activity of many mammalian genes, including apolipoprotein E (apoE), tyrosine aminotransferase (TAT), cytochrome P450 (P450), and the iron storage protein ferritin. The relative quantities of steady-state mRNA and protein in cells grown in the presence and absence of insulin are shown in the table below.

	− INSULIN		+ INSULIN	
	mRNA	**Protein**	**mRNA**	**Protein**
apoE	61	36	59	14
TAT	35	24	14	24
P450	24	62	12	31
ferritin	25	40	75	120

Assume that all differences in gene expression result from regulating either the rate of transcription or the rate of translation. Expression of which gene(s) is regulated by transcription? By translation?

Answer

Expression of TAT, P-450, and ferritin is regulated by transcription. Expression of apoE and TAT is regulated by translation.

16. Your laboratory investigates regulation of expression of *HR1*, which encodes a receptor protein for your favorite hormone. (Hormone receptors bind a specific hormone and mediate cellular responses to that hormone.) *HR1* is expressed in cultured mammalian cells. Your studies of these cells grown in the presence and absence of hormone have established the following facts concerning expression of *HR1*:

1. The rate at which the *HR1* gene is transcribed is 2-fold higher when cells are grown in the presence of hormone.
2. The half-life of *HR1* mRNA is 4-fold greater when cells are grown in the presence of hormone. (The half-life of a mRNA is the amount of time required to degrade half of the mRNA existing at any point in time.)
3. The quantity of *HR1* protein is 4-fold higher when cells are grown in the presence of hormone.
4. The half-life of *HR1* protein is 2-fold greater when cells are grown in the presence of hormone. (The half-life of a protein is the amount of time required to degrade half of the protein existing at any point in time.)

What are the relative quantities of *HR1* mRNA in cells grown in the presence and absence of hormone? What are the relative rates at which *HR1* mRNA is translated in cells grown in the presence and absence of hormone? Complete the following table.

	−Hormone	+Hormone
Relative quantity of *HR1* mRNA:	10	
Relative rate of translation of *HR1* mRNA:		10

Answer

	−Hormone	+Hormone
Relative quantity of *HR1* mRNA:	10	80
Relative rate of translation of *HR1* mRNA:	40	10

17. Transcription of *GAL1* (galactokinase) in yeast is regulated by *GAL4* and *GAL80*. GAL4 protein (abbreviated GAL4p) activates *GAL1* transcription by binding to UAS_G, an enhancer of *GAL1* expression. In the absence of galactose, GAL80p binds GAL4p, thereby inhibiting activation of *GAL1* transcription. In the presence of the inducer galactose, GAL80p does not bind or inhibit GAL4p. The quantity of galactokinase expressed in wild type, *gal4*, and *gal80* mutants is shown in the table below.

	GALACTOKINASE ACTIVITY	
Genotype	−Gal	+Gal
wild type	1	1,000
$gal4^-$	1	1
$gal80^-$	1,000	1,000

Gal = galactose

(a) What are the uninduced and induced levels of galactokinase in a $gal4^-$ $gal80^-$ double mutant?

(b) *Gal1-1* is an allele of *gal1* that eliminates function of UAS_G. What are the uninduced and induced levels of galactokinase in a *gal1-1* $gal80^-$ double mutant?

(c) *Gal4-2* is an allele of *gal4* that eliminates binding between GAL4p and GAL80p. *Gal4-2* does not alter binding of GAL4p to UAS_G. What are the uninduced and induced levels of galactokinase in a $gal4-2/GAL4^+$ diploid?

(d) *Gal80-3* is an allele of *gal80* in which the mutant GAL80p binds to and inhibits GAL4p even when galactose is present. What are the uninduced and induced levels of galactokinase in a $gal80-3/GAL80^+$ diploid?

(e) A diploid of genotype $gal4-2/GAL4^+$ $gal80-3/GAL80^+$ expresses 1 unit of galactokinase under both uninduced and induced conditions. Does the mutant GAL4p bind to the mutant GAL80p?

Answers

(a) uninduced = 1; induced =1

(b) uninduced = 1; induced = 1

(c) uninduced = 1,000; induced = 1,000

(d) uninduced = 1; induced = 1

(e) yes

18. Your laboratory investigates utilization of the hypothetical sugar "comatose" by yeast. As shown in the table on p. 253 (line 1), expression of comatose kinase (CK), an enzyme required for comatose degradation, is negatively regulated by glucose. You have isolated four mutations (*com-1, -2, -3,* and *-4*) that affect regulation of CK expression. Your genetic analysis indicates that *com-1* and *com-3* are linked to the CK structural gene, and that *com-2* and *com-4* are unlinked to each other and to the CK gene. The quantity of comatose kinase expressed by haploids and by heterozygous diploids is shown in the table below.

Comatose Kinase Enzyme Levels

	CARBON SOURCE		
Genotype	Glucose	Comatose	Glucose + Comatose
wild type	1	1,000	1
com-1	1	20	1
com-2	20	1,000	1,000
com-3	20	1,000	1,000
com-4	1	20	1
com-1/+	1	1,000	1
com-2/+	1	1,000	1
com-3/+	20	1,000	1,000
com-4/+	1	1,000	1

(a) One of the mutations affects an enhancer to which a positive transcription factor binds. The enhancer is located upstream of the CK gene and is required for induction of CK expression by comatose. The mutation destroys function of the enhancer. Which mutation is this?

(b) One of the mutations affects the *trans*-acting positive transcription factor discussed in part **(a)**. The mutation destroys function of the transcription factor. Which mutation is this?

(c) One of the mutations affects a silencer to which a negative transcription factor binds. The silencer is located upstream of the CK gene and is required for repression of CK expression by glucose. The mutation destroys function of the silencer. Which mutation is this?

(d) One of the mutations affects the *trans*-acting negative transcription factor discussed in part **(c)**. The mutation destroys function of the transcription factor. Which mutation is this?

➤ *Solutions*

(a) *Answer: com-1*

Enhancers are DNA regulatory sites that increase expression of a nearby gene. Enhancers exert their effect by binding positive transcription factors. Mutations that destroy an enhancer reduce gene expression by eliminating factor binding. Both *com-1* and *com-4* cause CK expression to be uninducible. *Com-4*, however, is unlinked to the CK gene, while *com-1* is linked. *com-1*, therefore, affects enhancer function. Because they affect regulatory sites in DNA, enhancer mutations are expected to be *cis*-dominant. It is not possible in this case, however, to establish whether *com-1* is *cis*-dominant. In a *com-1/+* heterozygote, the + allele is unaffected by mutation and, therefore, expressed normally. This prevents us from knowing whether the *com-1* mutant gene is inducible by comatose.

(b) *Answer: com-4*

Elimination of a *trans*-acting positive transcription factor by mutation causes the target gene to be unexpressed or expressed at lowered levels. Such mutations are almost always recessive. *Com-4* reduces CK gene expression, is recessive, and is unlinked to the CK structural gene. Thus, *com-4* destroys function of the positive transcription factor.

(c) *Answer: com-3*

Silencers are DNA regulatory sites that reduce expression of a nearby gene. Silencers exert their effect by binding negative transcription factors. Mutations that destroy a silencer increase gene expression by eliminating factor binding. Because they affect regulatory sites in DNA, silencer mutations are expected to be *cis*-dominant. Both *com-2* and *com-3* cause CK to be expressed at high levels even in the presence of glucose. *Com-2* is recessive and unlinked to CK, while *com-3* is dominant and linked. Thus, *com-3* likely affects silencer function.

(d) *Answer: com-2*

Elimination of a *trans*-acting negative transcription factor by mutation causes the target gene to be expressed at high levels or under inappropriate conditions. Such mutations are almost always recessive. This distinguishes them from other types of regulatory mutations that might cause elevated expression, such as operator constitutive mutations of bacteria or silencer mutations of eukaryotes. Both *com-2* and *com-3* cause CK to be expressed at high levels even in the presence of glucose. *Com-2* is recessive and unlinked to CK, while *com-3* is dominant and linked. Thus, *com-2* destroys function of the negative transcription factor.

Recombinant DNA Technology

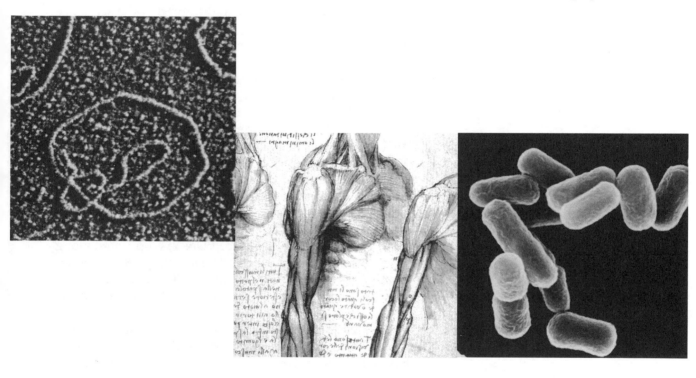

Summary

A single gene represents only a very small fraction of the genome of a complex eukaryotic organism, perhaps one in 100,000 different genes. The term **recombinant DNA,** which is often abbreviated as rDNA, refers to a variety of techniques that are used to isolate and manipulate individual genes. An isolated gene is amplified to produce millions of copies either *in vitro*, by enzymatic methods, or *in vivo*, as part of a small plasmid in bacteria. With large amounts of a single DNA molecule, it is possible to determine the sequence of the gene and to predict the sequence of the protein encoded by the gene. Recombinant DNA has led to enormous progress in understanding the structure, function, and regulation of genes. This knowledge has resulted in a multitude of applications, ranging from high resolution mapping of the human genome to the production of insect-resistant plants by transfer of bacterial toxin genes to plants.

Basic research led to the discovery of four tools which together were used to produce "recombinant" DNA. **Restriction enzymes** and **DNA ligase** permit splicing together two different DNA molecules. Bacterial viruses (**phage**) and **plasmids**, conferring new antibiotic resistances to common bacteria, were adapted as **vectors** to carry the recombinant DNA to bacteria. **Transformation** is used to introduce recombinant DNA molecules into host bacteria. The introduced DNA is replicated along with the chromosome of the host cell. The resulting cells or clones contain a single type of recombinant molecule. Finally, a process of screening is used to identify the clone containing a gene of interest.

Restriction enzymes were discovered by understanding the basis of host-controlled restriction and modification, a process by which bacteria "recognize" invading foreign DNA. Restriction enzymes are endonucleases that recognize a specific 4, 5, 6, or 8 base pair sequence within DNA. These sequences are usually palindromes; that is, the sequence reads the same on each strand of the DNA, e.g.:

5'ACGT3'
3'TGCA5'

Many different restriction enzymes have been isolated from bacteria. Each is named for the bacteria from which it was isolated. *Eco*RI, for example, is the first restriction enzyme identified for *Escherichia coli*. Restriction enzymes often make staggered cuts in the DNA, generating single base overhangs at the end of each fragment.

$$\text{5'-AGC}\overset{\downarrow}{\text{ T-3'}} \longrightarrow \text{5'-ACG} + \text{T-3'}$$
$$\text{3'-T}\underset{\uparrow}{\text{CG}}\text{A-5'} \qquad\quad \text{3'-T} \quad \text{GCA-5'}$$

If two different DNAs are cut with the same restriction enzyme and then mixed, the overhangs of different molecules, with complementary base sequences, may anneal to each other. These overhangs are sometimes referred to as **sticky** or **cohesive ends.** DNA **ligase** is used to reseal the phosphodiester backbone of the DNA molecules, generating the recombinant DNA.

Cloning utilizes **vectors**, which replicate *in vivo* and into which cloned DNA is inserted. A **plasmid** vector is a small circular double-stranded DNA molecule that contains an origin of DNA replication and a selectable marker, for example a gene coding for ampicillin resistance. Plasmid vectors often contain additional marker genes that identify insert-containing recombinant molecules. Recombinant plasmids are introduced into bacteria by transformation.

Phage have also been engineered to serve as recombinant DNA vectors. Phage vectors usually carry larger DNA inserts than plasmids. The recombinant phage infects a bacterium, replicates, and lyses its host, thereby releasing hundreds of copies of the recombinant molecule. Other types of cloning vectors include **cosmids** (hybrid between a phage and a plasmid), **YACS** (yeast artificial chromosomes) and **BACS** (bacterial artificial chromosomes).

A **library** refers to a collection of recombinant DNA clones that represent some portion of a complex genome. The size or number of individuals in the library can be used to calculate the probability that the library contains at least one copy of every single gene:

$$P = 1 - e^{N \ln(1-f)}$$

P = probability of including any single gene
N = number of clones in the library
f = fraction of the genome in each clone

Specialized libraries are often produced using the enzyme reverse transcriptase to make DNA copies of expressed mRNAs.

Such libraries are called **cDNA** libraries (for **complementary DNA**).

Screening is the process of identifying a clone of interest from a library or clone bank. This process requires some information about the gene which is being cloned: something that distinguishes this gene from other recombinant DNA molecules in the library. One screening method relies on DNA sequence similarity to a homologous gene previously cloned from another organism. Another screening method uses a vector that under the appropriate conditions will produce a protein from the recombinant DNA; the protein product may be identified with a functional assay, or with antibodies specific for the protein product of the gene.

Once a specific clone has been identified, a quick way to obtain information about the insert DNA is to prepare a **restriction map**. Purified clone DNA is cut with several different restriction endonucleases, and with pairs of restriction enzymes. Each restriction enzyme cut site is specific to the DNA sequence of the clone. The resulting fragments from each digest are separated by gel electrophoresis. The migration DNA fragments, stained with a fluorescent dye, are compared to the migration of standard molecular weight markers on the same gel. The overall size of the DNA clone is determined, restriction sites are ordered with respect to one another, and the distance between individual restriction sites is deduced.

In vitro methods for DNA manipulation utilize a wide variety of DNA-modifying enzymes. Each enzyme performs specific tasks, and geneticists use them to perform selective alterations of DNAs *in vitro*. For example, DNA polymerase I will replicate DNAs *in vitro* if provided a template, a primer, all four deoxyribonucleotides, and Mg^{2+}. A particularly important variation of this procedure, the **Polymerase Chain Reaction** (**PCR**) uses a thermostable DNA polymerase. During the PCR, the template molecule is denatured, copied, and denatured again in a single test tube for 20–30 rounds of DNA synthesis. This *in vitro* technique yields a geometric increase in the number of copies of the target gene. PCR has been adapted for many applications, including amplifying genes from ancient insects embedded in amber, diagnostic screens for the inheritance of mutant alleles, and various forensic purposes.

There are several methods to determine the sequence of bases in DNA. The most common methods are based on the dideoxyribonucleotide (ddNTP) chain terminator principle, developed by

Sir Frederick Sanger. The ddNTPs compete with dNTPs as substrates for DNA polymerase. Conventional methods for dideoxy sequencing use a segment of cloned single-stranded DNA as template in four sequencing reactions. Each reaction contains all four deoxyribonucleotides (one of them radiolabeled) and a small amount of one chain–terminating dideoxyribonucleotide. Newly synthesized DNA is elongated from the primer until the polymerase adds a ddNTP analog in place of the corresponding dNTP. The ddNTP is a chain terminator; it lacks a free 3'-OH so this DNA chain cannot be further elongated. Many template molecules are simultaneously copied in the sequencing of each reaction; chain termination, however, is a random event. The products of each sequencing reaction are a mixture of fragment sizes, all of which terminate at the same known base, and have the same 5' ends. The fragments are separated by size by gel electrophoresis, the gel is dried, and newly synthesized DNAs are detected by exposing the gel to X-ray film. Techniques for semi-automatic DNA sequencing using four fluorescent, distinctly labelled dideoxyribonucleotides as chain terminators increases the speed of DNA sequencing tremendously.

Recombinant DNA methods are frequently used to compare alleles of a gene. **Restriction fragment length polymorphisms (RFLPs)** are allelic differences that alter the restriction map of a region of DNA. Restriction maps of various alleles can be conveniently compared using **Southern blots** without having to clone each allele individually. In Southern blots, restriction fragments of total genomic DNA are separated by gel electrophoresis, transferred to a filter paper, and hybridized with radiolabel probe DNA. Only restriction fragments that share base sequence complementarity with the probe bind it tightly. Such fragments are visualized by exposing the filter to X-ray film. RFLPs alter the pattern of hybridizing bands because they alter the sizes of restriction fragments from the region.

Simple sequence length polymorphisms are increasingly important as genetic markers. Dinucleotides repeats (e.g., 5'CACACACACA3'; $(CA)_n$) occur in the genomic DNA of most organisms. The dinucleotide repeats or **microsatellite repeats** are dispersed over many chromosome sites. Dinucleotide repeats are often highly polymorphic between individuals as n (the number of repeats) varies for a particular chromosomal location of the repeat. These simple sequence-length polymorphisms are detected by PCR amplification, using primers flanking the dinucleotide repeats. Both RFLPs and microsatellite repeats are codominant genetic markers that are used in wide-ranging applications.

These include genetic mapping of disease genes in humans and agronomically valuable genes in plants, diagnosis of human disease and risk assessment, paternity testing, and forensic investigation of criminal suspects.

Self-Testing Questions

1. The sequence of the restriction site for the enzyme *Bam*HI is 5'G↓GATCC 3'.
   (a) What is the sequence of this restriction site on the complimentary DNA strand?
   (b) What is a common feature of restriction enzyme recognition sequences?
   (c) Predict how often the enzyme *Bam*HI will, on average, cut a plasmid molecule of 12,288 bp. Assume that all nucleotides are equally abundant.
   (d) How often, on average, will the enzyme *Hae*III (restriction site 5'GG↓CC3') cut this same molecule?

➤ *Solutions*
   (a) *Answer:* 5'G↓GATCC3'

   Use the base pairing rules for DNA to determine the sequence on the other strand. Remember that DNA sequences are commonly written 5' ⟶ 3'.

   (b) *Answer:* Most restriction enzyme recognition sites are **palindromes**.

   Most restriction sites are palindromes, that is they read the same on both strands. Read the *Bam*HI sequence from 5' to 3' on each strand.

   (c) *Answer:* On average, *Bam*HI is expected to cut this plasmid 3 times.

   The frequency that *Bam*HI is likely to cut a 12,288 bp plasmid is the product of the plasmid size multiplied by the probability that the bases of that recognition site are contiguous. There are 6 base pairs in this restriction site. The probability that a G occurs at the first position of any of the 6 bases in the plasmid is ¼. The probability that a G occurs at the second position is also ¼ . The probability that all six bases of the 6–base recognition sequence occur contiguously is (¼) × (¼) × (¼) × (¼) × (¼) × (¼) = ¼,096. In a plasmid of 12,288 base pairs, there are expected to be ^{12,288}⁄₄,₀₉₆ = 3 *Bam*HI recognition sites. The actual number of *Bam*HI restriction sites in the plasmid depends on its sequence.

(d) *Answer:* On average, the enzyme *Hae*III will cut this molecule 48 times.

*Hae*III cuts a 4–base pair recognition site. Therefore, the predicted number of restriction sites = 12,288 × (¼)^4 = 48.

2. Which of the following restriction enzymes will produce DNA fragments with complementary sticky ends?

*Eco*RI	5′G↓AATTC3′
*Bam*HI	5′G↓GATCC 3′
*Hpa*II	5′C↓GGC3′
*Bgl*II	5′A↓GATCT3′
*Sau*3A	5′↓GATC3′

Answer
The four base overhangs of *Bam*HI, *Bgl*II, and *Sau*3A are the same. Therefore, the ends of DNA fragments cut with any of these three enzymes will anneal to each other.

3. The sequence of a DNA molecule is shown below.

5′ATGCTTGACCTAATCGGCCGGAGAGGCTT3′
3′TACGAACTGGATTAGCCGGCCTCTCCGAA5′

(a) What molecules are produced if this DNA is digested by the restriction endonuclease *Hpa*II (specific for 5′ C↓GGC3′)?
(b) What molecules are produced if this DNA is digested by spleen phosphodiesterase, a 5′ exonuclease?

Answer
(a) 5′ATGCTTGACCTAATC3′   5′GGCCGGAGAGGCTT3′
    3′TACGAACTGGATTAGCC5′   3′GGCCTCTCCGAA5′

(b) If the reaction is allowed to go to completion, all four nucleotide 3′-monophosphates (dAMP, dTMP, dGMP, dCMP) will be produced as the DNA molecule is digested nucleotide-by-nucleotide from its ends.

4. The cloning vector pUC19 is ~ 2.7 kilobase pairs, confers ampicillin resistance, and has unique restriction sites for thirteen enzymes, including *Eco*RI, *Hind*III, and *Bam*HI, at the multiple cloning site. Use the data below to construct a restriction map for a new flavonoid glucosylase gene, *Bzteo*, which has been cloned into the plasmid pUC19. Which restriction enzyme was used to clone *Bzteo*?

Restriction enzyme digest	Fragment sizes (kbp)
*Eco*RI	3.0, 2.7
*Hind*III	2.7, 1.8, 1.2
*Bam*HI	2.7, 2.0, 1.0
*Eco*RI, *Hind*III	2.7, 1.8, 1.2
*Hind*III, *Bam*HI	2.7, 1.8, 1.0, 0.2

➤ *Solution*

Answer: The *Eco*RI restriction enzyme was probably used to clone *Bzteo*.

Plasmid clones have circular restriction maps. The sum of the fragment sizes from each digest should be identical. In this case the recombinant plasmid is 5.7 kbp. Look at the three single enzyme digests. Each produces a 2.7 kbp fragment, in this case corresponding to the original cloning vector. The *Eco*RI has a second fragment, 3 kbp in size, which corresponds to the size of the *Bzteo* clone insert. Since *Eco*RI does not cut within the insert, it is probably the restriction enzyme used for cloning.

*Hind*III and *Bam*HI each cut in the *Bzteo* segment: each enzyme produces three fragments from the recombinant plasmid, one fragment for the vector itself and two fragments from the insert. When both *Hind*III and *Bam*HI are used simultaneously, the 2.0 kbp *Bam*HI fragment is replaced by 1.8 and 0.2 kbp fragments. The 1.0 kbp *Bam*HI fragment remains unchanged. This places the *Hind*III site within the 2.0 kbp *Bam*HI fragment.

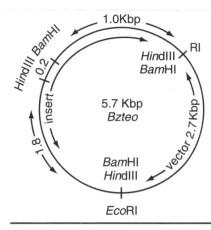

5. The restriction enzyme *Eco*RI recognizes the sequence 5′G↓AATTC3′ (the sequence is abbreviated so that only one strand is given.) In *E. coli* strain *K*, the third base of *Eco*RI recognition sites (adenine) is modified to N^6-methyladenine. A viral stock of bacteriophage λ is prepared in *E. coli* strain *B*, which lacks the *Eco*RI restriction/modification system. What happens when this λ is used to infect *E. coli K*?

Answer

The bacteriophage DNA is recognized by the *Eco*RI restriction endonuclease and cleaved into smaller subgenomic fragments; bacterial exonucleases then destroy these fragments.

6. You are asked to determine the restriction map of a recently cloned immunoglobulin μ gene. This gene was inserted in the polylinker site of a 2.6 kb plasmid vector. Sizes of various restriction fragments from the plasmid DNA are shown below. Derive a restriction map for the plasmid indicating the size of the inserted DNA and the position of the restriction sites in the insert. Assume that the plasmid vector has a single *Bam*HI site in the polylinker. The cloned fragment was isolated with the enzyme *Bam*HI.

Enzyme	Fragment sizes (kb)
*Eco*RV	5.4
*Bgl*I	2.1, 1.9, 1.4
*Pst*I	5.4
*Bam*HI	2.6, 2.8
*Eco*RV + *Bam*HI	2.6, 1.9, 0.9
*Eco*RV + *Bgl*I	2.1, 1.4, 1.3, 0.6
*Eco*RV + *Pst*I	3.6, 1.8
*Pst*I + *Bgl*I	1.9, 1.4, 1.2, 0.9

Answer

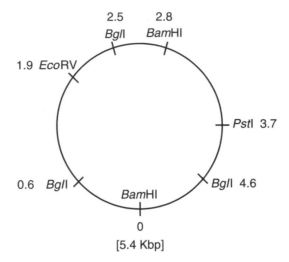

coordinates are in Kbp

7. A cDNA for blood clotting factor VIIIA, required by some hemophiliacs, is inserted into an *Eco*RI site of a 2.7kb plasmid cloning vector. The insert is 7.7 kb in length. The plasmid is amplified in the bacteria, and the insert is isolated from the plasmid. The insert is cut with several restriction enzymes. The resulting fragments are separated by agarose gel electrophoresis. The fragment sizes are listed below for each digest. Derive a restriction map for this factor VIIIA cDNA clone.

Enzyme	Fragment sizes
*Bam*HI:	2.7 kb, 5 kb
*Bgl*II:	4.3 kb, 3.4 kb
*Bam*HI + *Bgl*II:	3.4 kb, 2.7 kb, 1.6 kb

Hint: this is a linear restriction map.

Answer

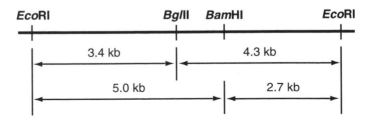

8. The first exon of the wild-type gene for β-globin is cut twice by the restriction enzyme *Dde*I (C↓TNAG, where N = any nucleotide), producing two fragments of 175 and 201 bp. The mutation causing sickle-cell anemia is a single base pair substitution CTGAT→ CTGTG of the β-globin gene. Amniocentesis is used to isolate cells and DNA from a young fetus; DNA is also isolated from its parents. The DNAs are cut with *Dde*I, separated on agarose gels, denatured and transferred to a nylon membrane. The DNAs are hybridized to a cloned probe corresponding to the first exon of the β-globin gene. Which of the RFLP profiles shown below indicates that the parents are both carriers of sickle-cell trait and that their fetus is homozygous for sickle-cell anemia?

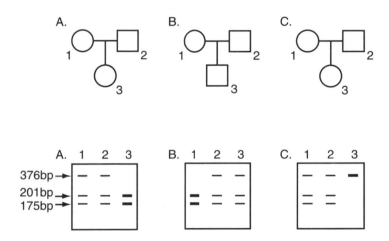

> **Solution**

Answer: In pedigree C, the fetus is homozygous for sickle-cell anemia, and the parents are heterozygous carriers.

The sickle-cell mutation eliminates one *Dde*I restriction site, so the hybridizing fragment from a sickle-cell globin gene is 376 bp in length. For each pedigree, what is the genotype of the parents? What is the genotype of the fetus? Only in pedigree C is the fetus homozygous for sickle cell trait and the parents heterozygous carriers.

9. cDNA clones are derived as DNA copies of expressed mRNA. What type of polymerase is required to make a single-stranded DNA copy of a pool of mRNA?

Answer Reverse transcriptase

10. Which of the following components is not required for the Polymerase Chain Reaction (PCR), an *in vitro* DNA synthesis method using thermostable DNA polymerase?
(a) template DNA
(b) dNTPs
(c) synthetic oligonucleotide primers
(d) RNA polymerase

Answer
(d) RNA polymerase synthesizes primers during *in vivo* DNA replication, but synthetic primers specific to the target gene are used during the PCR.

11. Oligonucleotide primers complementary to the *Bz* gene in maize amplify a 1.6 kbp fragment when wild-type genomic DNA is used as template for the PCR. These same primers amplify a ~2.0 kbp fragment using genomic DNA from the mutant allele *Bz-wm*. What type of mutation is *Bz-wm*?

Answer
Insertion mutation

12. The DNA sequence 3'TTCTGTCTGGATCTGAATC5' is a template for a dideoxy sequencing experiment. Which of the following oligonucleotides could be used as a primer for this sequencing experiment?
(a) 5' AAGACT3'
(b) 5' TTCTGT3'
(c) 5' AAGACA3'
(d) 5'GATTCA3'

Answer
(c) Only oligonucleotide (c) is fully complementary to the template strand. The polymerase will extend this primer in a 5'-to-3' direction.

13. If you want to synthesize radioactive DNA from a purified template with DNA polymerase, primer, and dNTPs, which of the following labelled nucleotides could you add to the reaction?

(a) $[\gamma - {}^{32}P]dATP$

(b) $[\alpha - {}^{32}P]ATP$

(c) $[\alpha - {}^{32}P]dATP$

➤ *Solution*

Answer: ^{32}P αdATP

Think about the structure of a deoxyribonucleotide triphosphate. The inner most phosphate group, designated α, is incorporated into the phosphodiester backbone of DNA. The outer phosphate groups, β and γ, are released as pyrophosphate when the dNTP is added to the new DNA. Answer **(b)** is incorrect because ribonucleotides are not substrates for DNA polymerase.

14. What are the sequences of the newly synthesized DNAs produced in sequencing reaction with DNA polymerase, the template 3′ CCTACGCGAATCATGCTTCA5′, the primer 5′GGATG3′, dATP, dCTP, dGTP, dTTP, and the dideoxyribonucleotide ddGTP?

Answer
5′GGATGCGCTTAGTACGAA**G**T3′
5′GGATGCGCTTAGTACGAA**G**3′
5′GGATGCGCTTAGTAC**G**3′
5′GGATGCGCTTA**G**3′
5′GGATGC**G**3′

Dideoxy-G residues are shown in boldface. Newly copied strands begin with the same 5′ base primer and end wherever a ddG has been added to that particular strand.

15. The autoradiogram from a dideoxy sequencing experiment is shown as a diagram below. The left set of four lanes represents the sequence of the wild-type allele. The right set of four lanes represents the sequence of the mutant allele. What is the DNA sequence deduced for the wild-type gene? The mutant gene? Indicate 5′ and 3′ ends of your deduced sequence. What is the relationship of the mutant to the wild-type gene?

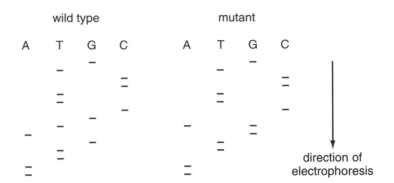

Answer

The deduced sequence of the wild-type gene is 5'AATTGATGCTTCCTG 3'. The deduced sequence of the mutant gene is 5'AATTGAGCTTCCTG 3'. The mutant is a one base pair deletion of the wild-type gene.

16. Two oligonucleotides (shown below) are mixed together and allowed to anneal. DNA polymerase, [α-^{32}P]-dATP, dGTP, dCTP, and ddTTP (dideoxyTTP) are added to the reaction.

Oligonucleotide #1:
5'-ATCGGGACGTTCCAGCGTGATCCATCCGTG-3'

Oligonucleotide #2:
5'-TTAGCCACGGATGGATCACGCT-3'

How many radiolabeled strands of DNA are produced? How long are they?

Answer

One DNA strand of 29 nucleotides is radiolabeled. A second DNA strand of 33 nucleotides is synthesized, but it is not radiolabeled.

17. The plasmid vector pBR322 confers resistance to the antibiotics ampicillin and tetracycline. The restriction enzyme *Bam*HI cuts once within this vector, within the tetracycline resistance gene. Vector and Drosophila DNA are cut with *Bam*HI, mixed and incubated with DNA ligase. If bacteria are transformed with this mixture, which antibiotic should be added to the agar plates to select bacteria with recombinant plasmids?

Answer

Ampicillin. Cloning into the *Bam*HI restriction site disrupts the tetracycline gene. Transformants containing recombinant plasmids are ampicillin resistant but tetracycline sensitive.

18. Cosmid vectors can be used to propagate recombinant DNAs as large as 35 kbp. The genome of the model organism *Arabidopsis thaliana* is 1×10^5 kbp. How many cosmids are needed in a library to insure that any single *Arabidopsis* gene has a 95% probability of being included in the library? Assume the average cosmid insert is 35 kbp.

➤ **Solution**

Answer: N = approximately 8,560 cosmid clones

For this problem, you need to consider the average size of the insert, the amount of DNA in the *Arabidopsis* genome, and a factor for the probability that a sequence is represented in the library. The equation relating these variables is:

$$P = 1 - e^{N \ln(1-f)},$$

where
N = number of clones in the library
P = probability of recovering any single clone
f = fraction of the genome in each clone

In this case, P = 0.95 and f = (35 kbp)/(10^5 kbp). Substituting those values and solving for N, we calculate that approximately 8,560 cosmid clones are required for 95% "coverage" of the *Arabidopsis* genome.

19. You are working with a clinical geneticist to determine why a young patient has a low level of hemoglobin, a condition called β^+ thalassemia.
(a) You collect a sample of the patient's DNA, and prepare a recombinant phage library with this DNA. Assuming you already have a clone of the wild-type allele, how will you screen this library?
(b) You have synthetic oligonucleotides which anneal to DNA sequences flanking the β-globin gene. How would

you use these oligonucleotides to isolate the patient's β-globin gene?

(c) How would you determine the molecular basis of this condition?

Answer

(a) Use the wild-type clone to prepare a hybridization probe for the phage library. Only DNA from those phage containing the patient's β-globin genes will hybridize to the probe.

(b) Isolate DNA from the patient and amplify the defective gene using the polymerase chain reaction.

(c) You might start by sequencing the isolated mutant allele. Comparing the sequence of the mutant to the wild-type allele may tell you if the mutation alters the promoter (thereby attenuating transcription), alters a splice site in the message, or otherwise interferes with gene expression.

20. Dideoxyinosine is used as a chemotherapeutic agent for individuals who are HIV positive. Suggest an explanation for why dideoxyinosine might slow virus propagation.

Answer

Dideoxyinosine is a nucleoside analog that acts as a chain terminator during DNA replication, thereby inhibiting replication of nucleic acids. By inhibiting viral replication, dideoxyinosine inhibits viral reproduction.

21. The metabolic protein lactate dehydrogenase (LDH) was purified from normal human muscle, and the purified protein was used to raise rabbit antibodies against LDH. A patient at the National Institutes of Health suffers from a generalized muscle weakness. A muscle biopsy was performed; this patient's LDH has an altered mobility during gel electrophoresis as compared to wild-type LDH. Your working hypothesis is that the patient's LDH has an altered amino acid sequence.

Total RNA was isolated from the patient's muscle biopsy.

(a) Describe how you would use the RNA to make a cDNA library.

(b) Describe how you would use the antibodies plus the cDNA library constructed in part **(a)** to isolate a cDNA clone corresponding to the patient's muscle LDH.

Answer

(a) First you would separate the mRNA from the total RNA; a convenient method is to pass the RNA through an affinity column consisting of the synthetic oligonucleotide poly(dT). Only mRNA has a poly A tail, and thus mRNA will preferentially bind to the column. Next, you elute the mRNA from the column, and use it as a template for cDNA synthesis. The synthetic oligonucleotide, oligo(dT), is used as a primer, and reverse transcriptase is used to copy the messenger RNA pool into a single strand of complementary DNA (cDNA). RNase H is used to digest the RNA template, and the second strand of DNA is synthesized with a DNA polymerase. The double-stranded cDNAs are then ligated into plasmid vectors that have been engineered to transcribe and translate inserted DNA from a nearby promoter. Bacteria are transformed with the recombinant molecules and spread on nutrient Petri plates. The resulting library constitutes a cDNA expression library, where each colony (or plaque) expresses one cDNA clone.

(b) A replica of the recombinant clones is made by layering a nylon membrane onto the agarose plate. Small amounts of proteins expressed from each clone adhere to the membranes. The membrane is treated with anti-LDH antibodies; the antibodies only bind to places on the membrane with LDH protein. Antibody binding is visualized with immunofluorescent techniques. The position of the reacting protein on the membrane replica is compared to the original petri plate to identify the cDNA clone for the patient's LDH.

22. Human clones "A" and "B" are linked to each other and to an inherited autosomal dominant disease. When used as hybridization probes on Southern blots, clones A and B detect several restriction fragment length polymorphisms.

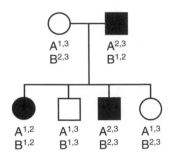

Inheritance of the disease and of polymorphic A and B alleles are shown in the diagram to the left. Which polymorphic allele is located on the same chromosome as the disease allele in this family?

Answer

Allele A^2 is coincident with inheritance of the disease in this family. All other alleles of either A or B are not coincident with the disease.

23. If you want to study the enhancers and promoters that control gene expression of a eukaryotic gene, would you choose a genomic DNA clone or a cDNA clone as starting material for your investigations?

➤ **Solution**

Answer: a genomic clone

A cDNA clone is prepared from mRNA, and has sequences corresponding to amino acid coding segments of a gene. A genomic clone is produced from nuclear DNA and, therefore, contains DNA sites that regulate gene expression but are not contained in the mRNA.

24. The insert of plasmid pCLM9 is a 1.2 kilobase *Bam*HI restriction fragment isolated from a sample of double-stranded mouse cDNA. Other than the *Bam*HI restriction sites at its termini, the insert of pCLM9 contains no other *Bam*HI restriction sites. When pCLM9 is used as a hybridization probe on a Southern blot of wild-type mouse genomic DNA digested with *Bam*HI, it hybridizes to two restriction fragments. One hybridizing fragment is 800 base pairs; the second is 1.5 kilobases. Assuming that the differences between the cDNA and genomic restriction fragments are due to the presence of introns within a gene, what can be said about the number and size of introns in this gene?

Answer

The gene contains at least one, but possibly more, introns in the region covered by the cDNA clone. Together, the sizes of the introns total at least 1.1 kilobases, but they could be considerably larger if more than one *Bam*HI site is present within the intron(s).

25. A template for a PCR reaction is shown below. Draw in the positions and polarities (5′ and 3′ ends) of the primers you would need to amplify gene c.

Answer

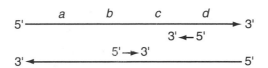

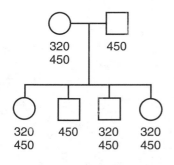

26. Depending on the individual tested, oligonucleotide primers complementary to a specific human gene amplify either a 320 bp, a 450 bp, or both 320 and 450 bp DNA fragments using the Polymerase Chain Reaction (PCR). These differing PCR products represent polymorphic alleles of a single gene. Inheritance of these alleles in one family is shown in the diagram to the left. Is the region amplified by these primers located on the X chromosome or on an autosome?

Answer

The region is located on an autosome.

topic 13
Population Genetics

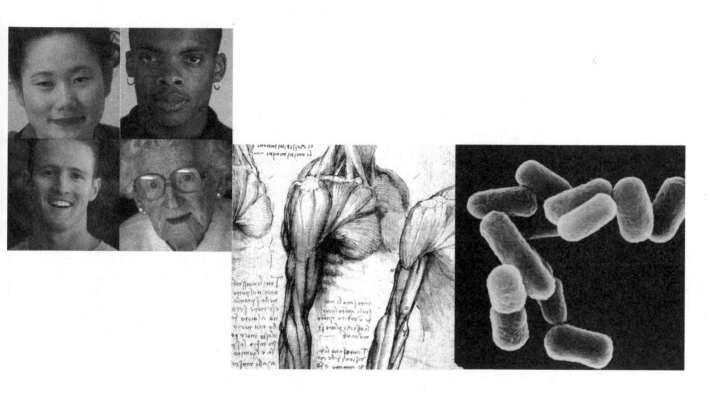

Summary

In sexually reproducing organisms the unit of evolutionary change in nature is a population composed of interbreeding individuals. **Population genetics** is the branch of genetics that seeks to explain the genetic structure of such populations, how this structure changes with time in response to various forces, and how these changes underlie the evolution of the population. In contrast with the laboratory situation, controlled matings are not possible in nature and rarely do we know the genotypes of all the individuals in a population. Nonetheless, by making a few simplifying assumptions, we can make mathematical models of natural populations to which we can apply the basic principles of Mendelian inheritance. A great deal of information about the genetic composition of populations and how this composition changes with time can be derived in this manner.

The sum total of all the genes within a particular interbreeding population constitutes the **gene pool** of that population. The genetic composition of a population is usually described in terms of **allele frequencies,** where the frequency of an allele is defined as the proportion of that particular allele among all alleles of a given gene in the gene pool. Much of population genetics is concerned with how allele frequencies change in response to various evolutionary forces and how rapidly these changes occur.

The **Hardy-Weinberg Law,** the fundamental principle of population genetics, states that in a large, randomly mating population, in which there is no mutation, selection, or migration, the allele frequencies remain constant from one generation to the next. Furthermore, the genotypic frequencies will be in the following proportions:

Genotype	Frequency
AA	p^2
Aa	$2pq$
aa	q^2

where p is the allele frequency of A, q is the allele frequency of a, and $p + q = 1$. Also note that $p^2 + 2pq + q^2 = 1$. When these relationships hold for a particular population, the population is said to be in **Hardy-Weinberg equilibrium** for the gene in question. This represents a static situation in which the allele frequencies and the genotype frequencies do not change in successive generations.

The most significant evolutionary force leading to change in the genetic composition of a population with time is **natural selection.** Selection occurs when there is differential **fitness** among genotypes; i.e., organisms of certain genotypes are better able to survive and reproduce than organisms of other genotypes. The most fit genotypes make a disproportionate contribution to the gene pool of the next generation with a consequent increase in frequency of the favored allele at the expense of the allele selected against. If selection is maintained, the favored allele will eventually replace the less favored allele. Such changes in genetic composition ultimately determine the evolutionary change of a population with time. Equations can be derived that precisely describe these changes in allele frequencies in successive generations.

In some cases, heterozygotes are the most fit genotype. When this occurs, selection does not lead to the elimination of either allele. Instead, both alleles are maintained at **stable equilibrium** values determined by the relative degree of selection against either homozygote. Thus, an allele that is deleterious in the homozygous condition (such as the one causing sickle cell anemia) can nonetheless be stably maintained in the population at reasonably high levels if it confers some benefit to heterozygotes (such as resistance to malaria in the case of the sickle cell allele). This is one mechanism by which selection can act to maintain, rather than eliminate, genetic variation in natural populations.

The gradual elimination of a deleterious allele from a population by selection is counterbalanced by the occurrence of new mutations to that allele. When the rate of elimination by selection is equal to the rate of new occurrences by mutation, the two opposing forces balance out and the allele frequency will remain constant at a stable equilibrium value determined by the relative values of the mutation rate and the strength of selection.

Self-Testing Questions

1. A sample of 500 people from the South Pacific Islands was tested for *MN* blood type with the following results:

Genotype	Number
MM	50
MN	200
NN	250

(a) What are the frequencies of the three different genotypes in this population?

(b) What are the allele frequencies of *M* and *N* in this population?

➤ **Solutions**

(a) *Answer:* frequency of *MM* = 10%; frequency of *MN* = 40%; frequency of *NN* = 50%

$$\begin{array}{l} 50 \ MM \\ 200 \ MN \\ \underline{250 \ NN} \\ 500 \ \text{Total} \end{array}$$

$$\text{frequency } MM = {}^{50}\!/_{500} = 10\%$$
$$MN = {}^{200}\!/_{500} = 40\%$$
$$NN = {}^{250}\!/_{500} = 50\%$$

(b) *Answer:* frequency of *M* = 30%; frequency of *N* = 70%

$$\begin{aligned} \text{frequency } M &= \text{frequency of homozygotes} + \tfrac{1}{2} \\ &\quad \text{frequency of heterozygotes} \\ &= 10\% + \tfrac{1}{2}(40\%) \\ &= 30\% \end{aligned}$$

$$\begin{aligned} \text{frequency } N &= \text{frequency of homozygotes} + \tfrac{1}{2} \\ &\quad \text{frequency of heterozygotes} \\ &= 50\% + \tfrac{1}{2}(40\%) \\ &= 70\% \\ &= 100\% - \text{frequency of } M \end{aligned}$$

Pitfalls. You cannot just assume that the population is in Hardy-Weinberg equilibrium.

2. In Drosophila there are two different alleles of the gene encoding alcohol dehydrogenase, Adh^F and Adh^S. A random sample of 150 flies is collected from a wild population and their genotypes for the *Adh* gene are determined. The following results are obtained:

Genotype	Number
Adh^F/Adh^F	18
Adh^F/Adh^S	36
Adh^S/Adh^S	96
	150 Total

(a) What are the frequencies of the three different genotypes in this population?

(b) What are the allele frequencies?

➤ *Solutions*

(a) *Answer:* frequency of Adh^F/Adh^F = 12%; frequency of Adh^F/Adh^S = 24%; frequency of Adh^S/Adh^S = 64%

$$18 \text{ F/F}$$
$$36 \text{ F/S}$$
$$\underline{96 \text{ S/S}}$$
$$150 \text{ Total}$$

frequency F/F = $^{18}/_{150}$ = 12%
frequency F/S = $^{36}/_{150}$ = 24%
frequency S/S = $^{96}/_{150}$ = 64%

(b) *Answer:* frequency of F = 24%; frequency of N = 76%

frequency F = 12% + ½(24%) = 24%
frequency N = 64% + ½(24%) = 76%

Pitfalls. You cannot just assume that the population is in Hardy-Weinberg equilibrium.

3. In another population of *Drosophila*, which is known to be in Hardy-Weinberg equilibrium for the *Adh* gene, the frequency of Adh^S homozygotes is 0.16.

(a) What is the frequency of the Adh^F allele in this population?

(b) If the population is made up of 2,000 flies, what is the total number of heterozygotes in this population?

➤ *Solutions*

(a) *Answer:* frequency of Adh^F = 0.6

If Adh^S/Adh^S = 0.16, then q, the allele frequency of Adh^S = $\sqrt{0.16}$ = 0.4. So p, the allele frequency of Adh^F = 1 − 0.4 = 0.6

(b) *Answer:* number of heterozygotes = 960

2,000 flies q = 0.4 p = 1.0 − 0.4 = 0.6
frequency of heterozygotes = $2pq$ = 2(0.6)(0.4) = .48
number of heterozygotes = (.48)(2,000) = 960

4. In a particular randomly mating population, approximately one person in 10,000 is an albino. Albinism is caused by homozygosity for an autosomal recessive allele.

   (a) What is the frequency of the allele for albinism in this population?

   (b) In this population how much more frequent are the heterozygotes for albinism than the homozygotes?

➤ *Solutions*

   (a) *Answer:* frequency of albino allele = .01

$$\text{albinos} = \frac{1}{10,000} = q^2$$

$$q = \text{frequency of albino allele}$$

$$= \sqrt{\frac{1}{10,000}}$$

$$= \frac{1}{100} \text{ or } .01$$

   (b) *Answer:* Heterozygotes for albinism are 198 times more frequent than homozygotes for albinism.

$$\text{allele } a = .01$$
$$\text{allele } A = .99$$
$$\text{heterozygotes} = 2(.01)(.99) = .0198$$
$$\text{heterozygotes/homozygotes} = \frac{.0198}{.0001} = 198$$

5. Shown below are the genotypic frequencies in two different populations for a particular gene with two different alleles:

	AA	**Aa**	**aa**
Population I	.09	.42	.49
Population II	.20	.20	.60

   (a) What are the allele frequencies of A and a in populations I and II?

   (b) Is either of these two populations in Hardy-Weinberg equilibrium? Defend your answer.

   (c) After 1 generation of random mating, what will be the genotypic frequencies in Population I?

(d) After 1 generation of random mating, what will be the allele frequencies in Population II?

(e) After 1 generation of random mating, what will be the genotypic frequencies in Population II?

➤ *Solutions*

(a) *Answer:* Population I: $A = .30$, $a = .70$; Population II: $A = .30$, $a = .70$

Population I
$A = .09 + \frac{1}{2}(.42) = .30$
$a = 1 - .30 = .70$

Population II
$A = .20 + \frac{1}{2}(.20) = .30$
$a = 1 - .30 = .70$

(b) *Answer:* Population I is in Hardy-Weinberg equilibrium; Population II is not.

Expected genotype frequencies for Hardy-Weinberg equilibrium with $p = .30$ and $q = .70$ are

$(.30)^2$ $AA = .09$
$2(.30)(.70)$ $Aa = .42$
$(.70)^2$ $aa = .49$

These frequencies match those of Population I but differ significantly from those in Population II. Therefore, Population I but not Population II is in Hardy-Weinberg equilibrium.

(c) *Answer:* The genotypic frequencies in Population I will be $AA = .09$; $Aa = .42$; and $aa = .49$.

After 1 generation of random mating, the genotypic frequencies in Population I will be

$AA = p^2 = (.30)^2 = .09$
$Aa = 2pq = 2(.30)(.70) = .42$
$aa = q^2 = (.70)^2 = .49$

i.e., unchanged from what they were.

(d) *Answer:* The allele frequencies in Population II will be $A = p = .30$ and $a = q = .70$.

After 1 generation of random mating, the allele frequencies in Population II remain the same as the starting frequencies:

$$p = .30$$
$$q = .70.$$

(e) *Answer:* The genotypic frequencies in Population II will be $AA = .09$; $Aa = .42$; and $aa = .49$.

With random mating, the genotype frequencies in Population II become the Hardy-Weinberg values:

$$AA = p^2 = (.30)^2 = .09$$
$$Aa = 2pq = 2(.30)(.70) = .42$$
$$aa = q^2 = (.70)^2 = .49.$$

---

6. Beginning with the general equation for selection:

$$q_{n+1} = \frac{q_n - sq_n^2}{1 - sq_n^2}$$

where

q_n = the allele frequency at generation n
q_{n+1} = the allele frequency at generation $n+1$
s = the selection coefficient,

derive the equation for selection against a recessive lethal or sterile allele:

$$q_{n+1} = \frac{q_n}{1 + q_n}$$

Answer

$$q_{n+1} = \frac{q_n - sq_n^2}{1 - sq_n^2}$$

for a lethal or sterile allele $s = 1$ so the equation simplifies to

$$q_{n+1} = \frac{(q_n - q_n^2)}{1 - q_n^2}$$

by factoring,

$$= \frac{q_n(1 - q_n)}{(1 + q_n)(1 - q_n)}$$

$$= \frac{q_n}{1 + q_n}$$

7. For selection against a recessive lethal or sterile allele show that after t generations of selection the allele frequency is

$$q_{n+t} = \frac{q_n}{(1 + tq_n)}$$

Answer

After 1 generation of complete selection

$$q_{n+1} = \frac{q_n}{1 + q_n}$$

After another generation of complete selection

$$q_{n+2} = \frac{q_{n+1}}{1 + q_{n+1}}$$

substituting $\dfrac{q_n}{1 + q_n}$ for q_{n+1} we get

$$q_{n+2} = \frac{\dfrac{q_n}{1 + q_n}}{1 + \dfrac{q_n}{1 + q_n}}$$

simplifying

$$= \frac{\dfrac{q_n}{1 + q_n}}{\dfrac{1 + q_n}{1 + q_n} + \dfrac{q_n}{1 + q_n}}$$

$$= \frac{\dfrac{q_n}{1 + q_n}}{\dfrac{1 + 2q_n}{1 + q_n}}$$

$$= \frac{q_n}{1 + 2q_n}$$

Similarly, it can be shown that after 3 generations of complete selection

$$q_{n+3} = \frac{q_n}{1 + 3q_n}$$

and, in general, after t generations of selection

$$q_{n+t} = \frac{q_n}{1 + tq_n}$$

8. Use the equations derived above to answer the following questions.

In chickens rr homozygotes have ragged feathers, while RR and Rr individuals have normal feathers. A poultry breeder finds that one in 400 chicks in his randomly mating flock are ragged. He removes all the affected chicks from the population before they mate and allows all the remaining birds to mate at random.

(a) What will be the allele frequency of r after 1 generation of selection?

(b) What will be the frequency of ragged chickens after 1 generation of selection?

(c) What will be the frequency of ragged chickens after 10 generations of selection?

(d) If the breeder wishes to reduce the allele frequency of r down to 25% of its initial frequency, for how many generations must he continue to impose complete selections against the ragged chicks?

➤ **Solutions**

(a) *Answer:* frequency of $r = \frac{1}{21}$ after 1 generation of selection

$$rr = \frac{1}{400} = q^2$$

so

$$q = \frac{1}{20} = \text{frequency of } r \text{ before selection}$$

$$q_{n+1} = \frac{\frac{1}{20}}{1 + \frac{1}{20}}$$

$$= \frac{1}{21} = \text{frequency of } r \text{ after 1 generation of selection}$$

(b) *Answer:* frequency of ragged chickens $= \frac{1}{441}$ after 1 generation of selection

$$\text{allele frequency} = \frac{1}{21} = q$$

$$\text{frequency of homozygous } rr = q^2$$

$$= \left(\frac{1}{21}\right)^2$$

$$= \frac{1}{441}$$

(c) *Answer:* frequency of ragged chickens = $\frac{1}{900}$ after 10 generations of selection

> After 10 generations of selection, the allele frequency of r becomes

$$q_{n+10} = \frac{\frac{1}{20}}{1 + 10\left(\frac{1}{20}\right)} = \frac{1}{30}$$

> Therefore, the frequency of affected chickens is

$$q^2 = \left(\frac{1}{30}\right)^2 = \frac{1}{900}$$

(d) *Answer:* 60 generations

> Rearranging the terms and simplifying the equation

$$q_{n+t} = \frac{q_n}{1 + tq_n}$$

> we get

$$t = \frac{1}{q_{n+t}} - \frac{1}{q_n}$$

> If the farmer wants to reduce the allele frequency of r to 25% of its initial value, we have

$$q_n = \frac{1}{20}$$

$$q_{n+t} = \frac{1}{20} \times \frac{1}{4} = \frac{1}{80}$$

> Therefore it will take

$$t = \frac{1}{\frac{1}{80}} - \frac{1}{\frac{1}{20}}$$
$$= 80 - 20$$
$$= 60 \text{ generations}$$

> of complete selection to reduce the allele frequency to this value.

9. In one large, randomly mating population of beetles, a variant spotted form is found. The variant is caused by homozygosity for a recessive allele, m.

Genotype	Frequency	Average Number of Progeny
MM	64%	200
Mm	32%	200
mm	4%	160

(a) What is the relative fitness of the heterozygotes?
(b) What is the relative fitness of *mm* homozygotes?
(c) What is the selection coefficient for *mm* homozygotes?
(d) What is the present allele frequency of *m*?
(e) What will be the allele frequency of *m* in the next generation?

➤ *Solutions*

(a) *Answer:* fitness of *Mm* = 1

Fitness of *Mm* heterozygotes = $^{200}/_{200}$ = 1.0

(b) *Answer:* fitness of *mm* = .80

Fitness of *mm* homozygotes = $^{160}/_{200}$ = .80 = *w*

(c) *Answer:* selection coefficient for *mm* = .20

$$w = 1 - s$$

where
w = fitness
s = selection coefficient

so

$$.80 = 1 - s$$
$$s = 1 - .80 = .20$$

(d) *Answer:* allele frequency of *m* = .20

present allele frequency of *m*

$$.04 + \tfrac{1}{2}(.32) = .04 + .16 = .20$$

(e) *Answer:* allele frequency of *m* = .194

$$q_{n+1} = \frac{\left(q_n - sq_n^2\right)}{\left(1 - sq_n^2\right)}$$

$q = .2$ $s = .2$

$$= \frac{\left(.2 - (.2 \times .04)\right)}{\left(1 - (.2 \times .04)\right)}$$

$$= \frac{.192}{.992}$$

$$= .194$$

10. Cystic fibrosis is the most common severe genetic disorder in people of northern European descent. It is caused by a recessive autosomal allele whose equilibrium frequency in the United States is about 2%. Until recently, affected individuals rarely survived to adulthood. Because the incidence of the disease allele is high for a lethal disorder, it has been suggested that, in the past, heterozygotes for cystic fibrosis had some selective advantage over normal homozygotes. If this idea is correct, what would the fitness of normal homozygotes have to be relative to heterozygotes to account for the present frequency of the cystic fibrosis allele?

➤ *Solution*

Answer: fitness of *CC* homozygotes = .98

for heterozygote advantage

$$\hat{q} = \frac{t}{s+t}$$

where

Genotype	Fitness
CC	$1 - t$
Cc	1
cc	$1 - s$

for cystic fibrosis
fitness = 0,
so $s = 1$
therefore at equilibrium:

$$.02 = \frac{t}{1+t}$$

$$(.02)(1 + t) = t$$
$$.02 + .02t = t$$
$$.02 = .98t$$

$$t = \frac{.02}{.98} = .02$$

therefore, fitness of *CC* homozygotes
$$= 1 - t = 1 - .02 = .98$$

---

11. In a certain population of butterflies, the fitnesses of genotypes A^1A^1, A^1A^2, and A^2A^2 are found to be 0.6, 1.0, and 0.8 respectively. The present allele frequency of allele A^2 is 0.5. When equilibrium is attained, what will be the allele frequencies of A^1 and A^2?

➤ **Solution**

Answer: allele frequency of A^1 = .33; allele frequency of A^2 = .67

Genotype	Fitness
A^1A^1	$0.6 = 1 - t$
A^1A^2	1.0
A^2A^2	$0.8 = 1 - s$

at equilibrium

$$\hat{q} = \frac{t}{s+t}$$

$s = 0.2 \quad t = 0.4$

$$\hat{q} = \frac{.4}{.2 + .4}$$
$$= .67$$
$$= \text{allele frequency of } A^2$$
$$\hat{p} = 1 - .67$$
$$= .33$$
$$= \text{allele frequency of } A^1$$

---

12. Achondroplastic dwarfism in humans is caused by a dominant allele, *D*. The frequency of affected individuals is 1.25×10^{-4}. Assume that this is an equilibrium value and that the frequency of homozygotes for *D* is negligibly small. Affected individuals are found to produce only 20% as many children as their unaffected brothers and sisters. From these data determine the mutation rate at which the normal recessive allele, *d*, mutates to the allele causing dwarfism.

➤ **Solution**

Answer: mutation rate = 5×10^{-5}

Affected individuals = homozygotes + heterozygotes for *D*
$$= q^2 + 2pq$$

At equilibrium, the frequency of affected people is 1.25×10^{-4}. Since the number of homozygotes is negligibly small, the affected individuals are all heterozygotes, so

$$2\hat{p}\hat{q} = 1.25 \times 10^{-4}$$

since $q \ll 1$

$$p \approx 1$$

so

$$2\hat{p}\hat{q} \approx 2\hat{q}$$

$$2\hat{q} \approx 1.25 \times 10^{-4}$$

$$\hat{q} \approx 6.25 \times 10^{-5}$$

For mutation-selection equilibrium for a dominant allele

$$\hat{q} = \frac{\mu}{s}$$

where s in this case $= 1 - .2 = 0.8$
thus

$$6.25 \times 10^{-5} = \frac{\mu}{.8}$$

$$\mu = \left(6.25 \times 10^{-5}\right)(0.8) = 5 \times 10^{-5}$$

---

13. The equilibrium allele frequency of a deleterious allele maintained by mutation is 0.02. The mutation rate for this allele is 4×10^{-5}.

(a) What is the value of the selection coefficient if the allele is recessive?

(b) What is the value of the selection coefficient if the allele is dominant?

(c) If the mutation rate doubles but the selection coefficient remains unchanged, what will be the new equilibrium frequency of the allele if it is recessive?

(d) If the mutation rate doubles but the selection coefficient remains unchanged, what will be the new equilibrium frequency of the allele if it is dominant?

➤ *Solutions*

(a) *Answer:* For a recessive allele, $s = .1$.

For mutation-selection equilibrium for a recessive allele

$$\hat{q} = \sqrt{\frac{\mu}{s}}$$

$\mu = 4 \times 10^{-5}$

$$.02 = \sqrt{\frac{4 \times 10^{-5}}{s}}$$

$$.0004 = \frac{4 \times 10^{-5}}{s}$$

$$.0004s = 4 \times 10^{-5}$$

$$s = \frac{4 \times 10^{-5}}{4 \times 10^{-4}} = .1$$

(b) *Answer:* For a dominant allele, $s = 2 \times 10^{-3}$.

For a dominant allele

$$\hat{q} = \frac{\mu}{s}$$

$$.02 = \frac{4 \times 10^{-5}}{s}$$

$$s = \frac{4 \times 10^{-5}}{2 \times 10^{-2}} = 2 \times 10^{-3}$$

(c) *Answer:* For a recessive allele, $\hat{q} = 2.83 \times 10^{-2}$.

$$\hat{q} = \sqrt{\frac{8 \times 10^{-5}}{.1}}$$

$$= \sqrt{8 \times 10^{-4}}$$

$$= 2.83 \times 10^{-2}$$

(d) *Answer:* For a dominant allele, $\hat{q} = 4 \times 10^{-2}$.

$$\hat{q} = \frac{\mu}{s}$$

$$\hat{q} = \frac{8 \times 10^{-5}}{2 \times 10^{-3}}$$

$$= 4 \times 10^{-2}$$

---

14. In some human populations, the frequency of Tay-Sachs disease is .0004. This disease is caused by homozygosity for a recessive allele and invariably causes fatality before four years of age. If the incidence of Tay-Sachs disease represents an equilibrium value maintained by a balance between mutation and selection, what is the mutation rate of the normal allele to the disease-causing allele?

➤ **Solution**

Answer: mutation rate = 4×10^{-4}

$$\hat{q} = \sqrt{\frac{\mu}{s}}$$

if the disease frequency $\hat{q}^2 = .0004$ then $\hat{q} = .02$

$$.02 = \sqrt{\frac{\mu}{s}}$$

where $s = 1$ for a lethal disorder

$$.02 = \sqrt{\mu}$$

$$\mu = (.02)^2$$

$$\mu = 4 \times 10^{-4}$$

---

15. In a population of guppies, an albino form is caused by a recessive allele, *a*. Experiments indicate that *AA* and *Aa* genotypes produce an average of 150 offspring each, whereas *aa* guppies produce an average of only 60 offspring. If *A* mutates to *a* at a rate of 2.4×10^{-6}, what is the expected frequency of the *a* allele in the population at equilibrium?

➤ **Solution**

Answer: expected frequency of $a = 2 \times 10^{-3}$

$$\hat{q} = \sqrt{\frac{\mu}{s}}$$

$$\mu = 2.4 \times 10^{-6}$$

$$s = 1 - \text{fitness}$$

$$= 1 - \frac{60}{150}$$

$$= 1 - .4$$

$$= .6$$

$$\hat{q} = \sqrt{\frac{2.4 \times 10^{-6}}{.6}}$$

$$\hat{q} = \sqrt{4 \times 10^{-6}} = 2 \times 10^{-3}$$

16. Assume that extreme nearsightedness is controlled by a recessive allele, *n*. Assume that in prehistoric times, people with this condition invariably died in early childhood because of the inability to see some lurking danger.
    (a) If the mutation rate is 4×10^{-5}, what would the allele frequency of *n* have been at equilibrium?
    (b) With the introduction of eyeglasses, the survival of affected individuals is increased to 90% of normal. What will be the new equilibrium frequency of *n* under these conditions, if the mutation rate remains constant?

➤ **Solutions**

(a) *Answer:* $\hat{q}$ of $n = 6.32 \times 10^{-3}$

$$\hat{q} = \sqrt{\frac{\mu}{s}}$$

$\mu = 4 \times 10 - 5 \quad s = 1$

$$\hat{q} = \sqrt{\mu}$$

$$= \sqrt{4 \times 10^{-5}}$$

$$= 6.32 \times 10^{-3}$$

(b) *Answer:* $\hat{q}$ of $n = 2 \times 10^{-2}$

$$\hat{q} = \sqrt{\frac{\mu}{s}}$$

fitness now $= .9 = 1 - s$
so $s = .1$

$$\hat{q} = \sqrt{\frac{4 \times 10^{-5}}{.1}}$$

$$= \sqrt{4 \times 10^{-4}}$$

$$= 2 \times 10^{-2}$$

glossary

Acentric: A chromosome or chromosome fragment that is lacking a centromere.

Acrocentric: A chromosome in which the centromere is located close to an end.

Adjacent Segregation: The meiotic segregation pattern in a translocation heterozygote that results in the movement of one standard and one translocated chromosome to each of the two poles.

Allele: One of the two or more alternative forms of a given gene. Alleles differ from each other in their precise DNA sequence and in the phenotypes they confer.

Allotetraploid: A polyploid containing two complete diploid sets of chromosomes, each derived from a different species.

Alternate Segregation: The meiotic segregation pattern in a translocation heterozygote that results in the movement of both standard chromosomes to one pole and both translocated chromosomes to the other pole.

Alternative Splicing: The generation of different mature mRNAs from a single eukaryotic gene by splicing together different sets of exons derived from the same primary transcript.

Aminoacyl Synthetase: One of a set of twenty different enzymes that catalyze the attachment of a specific amino acid to its cognate tRNA.

Anaphase: The stage of mitosis or meiosis at which sister chromatids (in mitosis or meiosis II) or homologous chromosomes (in meiosis I) separate and move to opposite poles.

Aneuploid: A cell or organism with a chromosome complement that differs from the normal chromosome complement of that species.

Aneuploidy: The condition of having an unbalanced genetic complement: i.e., a chromosome number that is not an exact integral multiple of the euploid number. Aneuploidy sometimes refers to any gross genetic imbalance, such as that associated with a large deletion or duplication.

Anticodon: A sequence of three nucleotides in a tRNA molecule that is complementary to and pairs with an mRNA codon that specifies the amino acid carried by that tRNA.

Ascospore: In fungi, one of the haploid spores produced following meiosis. An ascospore is enclosed within an ascus and represents the offspring of a fungal cross.

Ascus: In fungi, the saclike structure that contains all of the haploid spores resulting from the meiotic divisions of one cell.

Attenuation: A mechanism of regulation in which transcription of an operon that encodes enzymes involved in amino acid biosynthesis is terminated prematurely if available levels of the corresponding amino acid are sufficiently high.

Attenuator: A regulatory sequence of DNA in an operon that governs whether transcription proceeds to the structural genes based on the availability of the corresponding amino acid. The attenuator is located between the operator and the structural genes of certain bacterial operons involved in amino acid biosynthesis.

Autosome: Any of the chromosomes for a given species other than the sex chromosomes. The content of autosomal chromosomes and the pattern of inheritance of autosomal genes are the same for both sexes of a species.

Autotetraploid: A polyploid containing four haploid sets of chromosomes, all derived from the same species.

Auxotroph: A nutritional mutant in bacteria or fungi that can only grow on an appropriately supplemented medium.

Barr Body: The condensed mass of chromatin in the interphase nuclei of female mammals that corresponds to an inactivated X chromosome.

Bivalent: A synapsed pair of homologues during the first meiotic prophase. Because each chromosome consists of two sister chromatids at this stage, a bivalent is also referred to as a tetrad.

Branch Migration: The breakage and reformation of hydrogen bonds by which one DNA strand displaces a second strand when both are competing for pairing with the same complementary strand. During recombination, the site of connection between two heteroduplex DNA molecules can move in a zipperlike fashion via branch migration from the point of initial exchange.

Burst Size: The number of progeny phage particles released upon lysis of an infected host cell.

Cap: A posttranscriptional modification of the 5′ end of a eukaryotic mRNA, which involves the addition of a guanine residue by a 5′-5′ linkage (rather than the usual 3′-5′ phosphodiester bond) and the methylation of the purine ring.

Catabolite Repression: A regulatory mechanism in bacteria that prevents the induction of certain operons involved in the metabolism of sugars other than glucose (e.g., lactose or arabinose) when glucose, the preferred metabolite, is available in the growth medium.

cDNA: "Complementary" DNA. A DNA copy of an RNA molecule. cDNA is synthesized in vitro using, in part, the enzyme reverse transcriptase. cDNAs are particularly valuable for eukaryotic genes that contain introns, because they contain exon sequences joined together as in the mature mRNA.

Centimorgan: The basic unit in which recombination frequency is measured. One centimorgan is equal to a recombination frequency of one percent and is equivalent to one map unit.

Centromere: The specialized region of a chromosome necessary for proper chromosome movement during mitosis and meiosis. Centromeres are the sites at which spindle fibers attach to chromosomes.

Chain-Terminating Nucleotide: A 2′,3′-dideoxynucleotide triphosphate that, when incorporated into a growing DNA strand, causes elongation to terminate at that point. The modified chain-terminating nucleotide lacks a 3′ OH group and is thus unable to form a phosphodiester bond with the next incoming nucleotide triphosphate.

Chromatid: One of the two identical copies of a chromosome that remain attached at the centromere after chromosome replication in mitotic or meiotic prophase. After the centromere divides, the sister chromatids become separate chromosomes. Each chromatid is a single molecule of DNA.

Chromatin: The fundamental substance of which chromosomes are composed, consisting primarily of DNA and histones.

Chromosome: A threadlike structure, composed primarily of DNA and protein, that contains a specific set of coded instructions arranged as a linear series of genes. Each eukaryotic species has its own characteristic number of chromosomes located in the nucleus.

Cis-Dominant: The condition in which the phenotypic manifestation of an allele in a diploid or partial diploid is limited to those structural genes physically linked to that allele (e.g., mutations affecting promoters or operators).

Clone: A group of identical cells derived from a single common progenitor. Also, a collection of identical replicates of a given fragment of DNA generated by recombinant DNA techniques. Also used to refer to the process by which such DNA replicates are produced.

Co-repressor: A small effector molecule or ligand that binds to the repressor protein, thereby enabling it to bind to the operator to block transcription.

Codominance: A type of inheritance in which both alleles in a heterozygote are expressed equally and fully.

Codon: One of the sixty-four possible nucleotide triplets in mRNA that encodes a particular amino acid or that signals the end of translation.

Coefficient of Coincidence: A measure of how closely the observed number of double crossovers in two adjacent intervals matches the expected number if the events were independent. The coefficient of coincidence is equal to the observed number of double crossovers divided by the expected number; it is also equal to 1 minus the interference.

Colchicine: An alkaloid drug that blocks the assembly of spindle fibers and thus prevents the normal movement of chromosomes during mitosis or meiosis. Colchicine is used to artificially produce polyploids.

Complete Medium: A richly supplemented culture medium that provides all of the nutrients essential for the growth and reproduction of all strains of fungi, bacteria, or other microorganisms, even nutritional mutants whose biosynthetic capacity is impaired.

Conjugation: The process by which DNA is transferred from a donor cell that contains a fertility factor to a recipient cell that does not contain a fertility factor through a close union of the two cells.

Constitutive: The constant, unregulated expression of a gene or set of genes. Constitutive expression of regulated genes can be caused by mutations affecting regulatory components or by the absence of an appropriate effector molecule.

Crossing-Over: The process in which homologous chromosomes physically exchange corresponding segments during meiotic prophase by a mechanism involving the breakage and reunion of chromosomes. Crossing-over is the physical basis of recombination.

Cytokinesis: In cell division, the separation or partitioning of the cytoplasmic portion of the cell. Cytokinesis usually accompanies the mitotic or meiotic division of the nucleus.

Deletion: A chromosome aberration in which a chromosome segment is entirely lacking; also called a deficiency.

Dicentric: A chromosome that contains two discrete centromeres.

Dihybrid Cross: A mating in which both parents are heterozygous for two specified pairs of alleles.

Diploid: The state of having two complete homologous sets of chromosomes (with the possible exception of sex chromosomes) in each somatic cell nucleus.

Diplotene: A substage of meiotic prophase I at which the highly condensed paired homologues begin to repel each other and are held togeth-er only at the cytologically visible chiasmata.

DNA Polymerase: Any of several different enzymes that catalyze the formation of phosphodiester bonds between deoxyribonucleotide triphosphates aligned along a template DNA strand, thereby mediating the synthesis of DNA.

DNA Polymerase I: In *E. coli*, the DNA polymerase that completes the synthesis of Okazaki fragments. When two Okazaki fragments abut, DNA polymerase I digests the RNA primer of the preceding Okazaki fragment and replaces it with DNA. The adjoining fragments are then sealed by DNA ligase.

DNA Polymerase III: In *E. coli*, the enzyme that plays the primary role in DNA replication. DNA polymerase III continuously extends the leading strand. On the lagging strand, it elongates each new Okazaki fragment, beginning from an RNA primer, until it reaches the RNA primer of the preceding Okazaki fragment.

Dominant: The classification of an allele whose phenotypic effect is manifested in the heterozygous condition.

Duplication: A chromosome aberration in which an extra copy of a chromosome segment is present.

Enhancer: A regulatory sequence of DNA in eukaryotes that controls the abundance, tissue specificity, and temporal pattern of transcription of an associated gene. The effect of an enhancer is mediated through the sequence-specific binding of regulatory proteins that stimulate the activity of RNA polymerase bound at the promoter.

Euploidy: The condition of having a balanced set of chromosomes: i.e., an exact integral multiple of the haploid set of chromosomes.

Exon: A segment of a eukaryotic gene present in the mature mRNA molecule after the primary transcript is spliced to remove introns. Exons contain all of the protein-coding sequences of a gene as well as the untranslated sequences located at the 5′ and 3′ ends of the eukaryotic mRNA.

Ferritin: An intracellular iron storage protein.

Fertility Factor (F): A genetic element of bacteria whose presence in cells confers the ability to mate with cells not containing a fertility factor. The fertility factor can exist as an independent plasmid (an F⁺ cell), as part of a plasmid containing both bacterial and fertility factor DNA (an F′ cell), or as part of the bacterial chromosome (an Hfr cell).

First-Division Segregation: The segregation of a heterozygous allele pair into separate nuclei at the first meiotic division. In ordered tetrads, first-division segregation is indicated by a 4:4 spore pattern in which all spores in one half of the ascus contain one allele and all spores in the other half contain the other allele.

Fitness: A measure of the relative ability of individuals of a given genotype to survive and reproduce compared with that of the most fit genotype in a population. The most fit genotype is assigned a value of 1.

Frameshift Mutation: A mutation caused by the insertion or deletion of a single nucleotide in the coding sequence of a gene. The result is the production of an mRNA in which the triplet sequence of codons is read out of frame from the site of mutation onward by the translational machinery.

Gamete: A reproductive or germ cell, such as a sperm or an egg, that is produced following meiosis. Gametes contain a haploid number of chromosomes.

Gene: The fundamental unit of inheritance. A gene occupies a specific location on a chromosome and corresponds to a segment of DNA that in most cases encodes a single polypeptide.

Genetic Code: The molecular rules specifying the relationship between a nucleotide sequence in DNA or RNA and the corresponding amino acid sequence in a polypeptide. Also, the set of sixty-four triplet codons in mRNA and the corresponding amino acids (or termination signals) they specify.

Genotype: The genetic constitution of an organism with respect to the particular gene or genes under consideration.

Hemizygous: A condition in diploid organisms in which only a single allele is present rather than the usual two alleles. Examples include all X-linked genes of XY males.

Heterochromatin: Dark-staining, highly condensed regions of a chromosome that persist in the condensed state throughout the cell cycle. Unexpressed chromosomes (e.g., the Barr body) and regions of chromosomes containing few genes are often found as heterochromatin.

Heteroduplex: A DNA molecule containing segments in which the complementary paired strands are initially derived from two different parental molecules. Heteroduplex molecules are most commonly generated as an intermediate step in recombination.

Heterogametic: In organisms in which the sexes differ in sex chromosome makeup, heterogametic refers to the sex in which the two sex chromosomes in the pair are dissimilar. Examples include XY males in Drosophila and humans.

Heterozygote: A diploid individual that carries two different alleles for a given gene.

Histones: Proteins that physically associate with DNA in chromatin to form the physical structure of chromosomes. Histones contain a high proportion of the basic amino acids lysine and arginine, and their resulting positive charge neutralizes the negatively charged DNA.

Holliday Intermediate: The X-shaped structure formed after single-strand exchange between two DNA molecules. The Holliday intermediate occurs as an intermediate step in the molecular mechanism of recombination.

Homogametic: In organisms in which the sexes differ in sex chromosome makeup, homogametic refers to the sex in which the two sex chromosomes in the pair are alike. Examples include XX females in Drosophila and humans.

Homologue: In a diploid organism, one of the two members of a pair of chromosomes that share the same physical appearance, contain equivalent genes, and synapse and segregate from each other in meiosis. Also called a homologous chromosome.

Homozygote: A diploid individual that carries two copies of the same allele for a given gene.

In Utero: Inside the uterus. In utero is used in reference to diagnostic tests performed on a fetus prior to birth.

Incomplete Dominance: A type of inheritance in which the phenotype of a heterozygote is intermediate in appearance between the phenotypes of the respective homozygotes.

Induced: The state in which the expression of a particular gene or set of genes is activated by either the inactivation of a negative regulatory factor or the activation of a positive regulatory factor.

Inducer: A small effector molecule that stimulates the transcriptional activity of a particular bacterial operon by binding to and inactivating a repressor protein.

Interference: A measure of the effect of one crossover event on the occurrence of a second crossover event in an adjacent chromosome interval. Because of interference, the frequency of a double crossover is usually less than would be expected if the two events were independent. Interference is equal to 1 minus the coefficient of coincidence.

Interphase: The part of the cell cycle composed of the period during which DNA replication occurs (S phase) and the two gap periods (G2 and G1) that precede and follow it. Sometimes referred to, misleadingly, as a "resting stage."

Intron: An intervening or noncoding sequence of DNA that separates the protein-coding segments of eukaryotic genes into discrete blocks (exons). Both introns and exons are transcribed, but the intron sequences are removed in nuclear splicing reactions and are therefore not represented in the mature mRNA exported to the cytoplasm for translation.

Inversion: A chromosome aberration in which a chromosome segment is broken and reattached after being rotated 180 degrees. An inversion causes a reversal of the gene sequence for that segment relative to the rest of the chromosome.

Karyotype: The complete chromosome complement of an individual. A karyotype often specifically refers to a photographic display of the chromosomes, which have been aligned in orderly fashion according to size and centromere position.

Lagging Strand: In DNA replication, the daughter strand that is being synthesized discontinuously as a series of short Okazaki fragments in the opposite direction from the advancing replication fork.

Leader Sequence: The initial segment of the mRNA molecule, from the 5' end to the start codon of the first structural gene, within which the attenuator is located. The leader sequence is present only in bacterial operons subject to attenuation.

Leading Strand: In DNA replication, the daughter strand that is being synthesized continuously in the same direction that the replication fork is advancing.

Library: A collection of cloned fragments of DNA, which together include the entire genome of a given species (a genomic library) or represent all of the mRNA transcripts made in a particular type of cell or tissue (a cDNA library).

Ligand: A small effector molecule, such as an inducer or co-repressor, that acts by binding to a specific site on a particular target protein.

Ligase: An enzyme that catalyzes the formation of a phosphodiester bond between the 3'-OH and a 5'-P of adjacent nucleotides. In DNA replication and repair, DNA ligase seals adjoining polynucleotide strands in a continuous molecule.

Linkage: The condition in which two nonallelic genes tend to be inherited together, rather than independently, because they are physically located on the same chromosome.

Linkage Map: A diagram showing the linear order of genes located on the same chromosome and the relative distances between them in recombinational units.

Linked: The condition in which a pair of nonallelic genes fails to undergo independent assortment because they are physically located on the same chromosome.

Locus: The position along a chromosome where a particular gene is physically located. A locus is sometimes used to refer to the gene occupying that position (e.g., the white locus).

Lysate: The population of phage particles released from host bacteria upon lysis.

Lysogen: A bacterial strain that harbors a prophage. Such strains may undergo lysis, with the release of progeny phage particles, under conditions that favor the excision of the phage chromosome from the bacterial chromosome and the activation of its genes.

Lysogeny: One of the two reproductive pathways available to a temperate phage upon infecting a host cell. In lysogeny, the phage chromosome integrates into and replicates as part of the bacterial chromosome. Under these conditions, the phage genes are not expressed and progeny phage are not produced.

Map Unit: The basic unit in which recombination frequency is measured. One map unit is equal to a recombination frequency of one percent and is equivalent to one Centimorgan.

Maternal: The portion of an individual's genetic composition that is derived from the female parent.

Metacentric: A chromosome in which the centromere is located close to the middle.

Metaphase: The stage of mitosis and meiosis at which chromosomes are aligned in the middle of the dividing cell and spindle fibers have attached to the centromeres.

Minimal Medium: The culture medium that provides the smallest set of nutrients required to permit the growth and reproduction of wild-type strains of fungi, bacteria, or other microorganisms.

Mitosis: The part of the cell cycle during which nuclear division occurs and replicated copies of each chromosome are distributed to each daughter nucleus. Mitosis is composed of the events of prophase, metaphase, anaphase, and telophase.

Monohybrid Cross: A mating in which both parents are heterozygous for a specified pair of alleles.

Monoploid Number: The basic number of chromosomes in a polyploid series.

Monosomic: The condition in a normally diploid cell or organism in which one chromosome is present in one copy rather than the normal two copies. Monosomic cells or individuals have a chromosome content of 2N−1.

Monosomy: The condition in which one member of a chromosome pair is completely missing from an otherwise diploid organism.

Multiple Alleles: The existence of more than two different mutational forms of a given gene in a population of organisms.

Mutation: A sudden, heritable alteration in a gene from one allelic form to another. Mutations may occur spontaneously or, at higher frequencies, after exposure to certain physical or chemical agents that damage DNA.

Negative Regulation: A mode of regulation in which expression of a gene or set of genes is prevented by the activity of a specific regulatory protein. In the absence of a negative regulatory protein, the genes are expressed.

Nondisjunction: An error in nuclear division in which one or more chromosomes fail to segregate properly into daughter nuclei. Nondisjunction can occur during meiosis or mitosis.

Nonparental Ditype: A spore pattern in a fungal tetrad in which only two genotypic classes of spore are found containing allelic combinations not present in the parents. If the parents are genotypes *AB* and *ab*, a nonparental ditype ascus would contain two *Ab* spores and two *aB* spores.

Obligate Heterozygote: An individual in a pedigree that must be heterozygous for the gene in question on the basis of information provided in a pedigree.

Octad: The linear array of eight spores within an ascus produced after the mitotic division of the four haploid products of meiosis.

Okazaki Fragment: A small, newly synthesized strand of DNA, attached to an RNA primer, that is one of the intermediates in the discontinuous synthesis of the lagging strand during DNA replication. Okazaki fragments are ligated together to generate the complete and contiguous lagging strand.

Oligonucleotide Primer: A short, single-stranded segment of synthetic DNA that serves as a primer for the in vitro enzymatic synthesis of new DNA strands.

Operator: A DNA segment in an operon that controls the transcription of an adjacent set of genes by its interactions with specific regulatory proteins that recognize and bind to it.

Operon: A set of adjacent protein-coding genes whose expression is coordinately regulated by transcription into a polycistronic mRNA, plus the promoter and operator that control these genes. Many bacterial genes are organized into operons.

Ordered Tetrad: All the products of a single meiosis in an ascus recovered in a linear order that exactly reveals the pattern of chromosome segregation in the two meiotic divisions.

Paracentric: An inversion in which the inverted segment does not include the centromere.

Parental: A progeny cell or offspring carrying a chromosome that has not crossed over between the genes in question and therefore carries the same combination of alleles originally present in the parent.

Paternal: The portion of an individual's genetic composition that is derived from the male parent.

Pedigree: A diagram representing the familial relationships and transmission of a gene or trait over several generations. Pedigrees are most commonly used in human genetics.

Penetrance: A measure of the phenotypic expression associated with a given genotype. Penetrance is defined as the proportion of individuals of the specified genotype that manifest the corresponding phenotype under a given set of environmental conditions.

Peptide Bond: The covalent bond between amino acids in a polypeptide chain. Peptide bonds link the amino group of one amino acid to the carboxyl group of the next.

Pericentric: An inversion in which the centromere is included in the inverted segment.

Phage: A virus that infects bacteria and is dependent on the bacterial host for reproduction; also called bacteriophage.

Phenotype: The physical appearance of an organism with respect to the particular trait or traits under consideration. Phenotypes are conferred by the organism's genotype.

Plaque-forming Units: The number of phage particles in a given sample as determined by the ability of each individual phage to infect a host cell in a bacterial lawn and eventually give rise to a plaque.

Plasmid: A small, circular, nonessential, extra-chromosomal, self-replicating genetic element present in many bacterial species. Plasmids often carry genes that confer antibiotic resistance and can be transmitted from cell to cell via conjugation.

Plasmid Vector: A small, circular molecule of DNA containing an origin of replication, an antibiotic resistance gene, and one or more restriction enzyme cleavage sites where a foreign DNA fragment can be inserted for cloning in a bacterial host. Plasmid vectors are isolated from bacteria.

Poly(A) Tail: A long sequence of adenine residues added as a post-transcriptional modification of the 3' end of most eukaryotic mRNAs.

Polycistronic mRNA: An mRNA molecule that contains the coding information from two or more adjacent genes. Polycistronic mRNA is translated to yield many different polypeptides.

Polynucleotide: A linear polymer consisting of a sequence of nucleotides linked by phosphodiester bonds.

Polypeptide: A linear polymer consisting of a sequence of amino acids linked by peptide bonds.

Polyploidy: The condition of having a chromosome complement consisting of three or more complete haploid sets.

Polytene: Refers to the giant chromosomes in Drosophila and some other organisms that are formed when the chromosomes undergo many rounds of replication without nuclear division. Each chromosome consists of many identical chromatids precisely aligned along their length.

Positive Regulation: A mode of regulation in which expression of a gene or set of genes is activated by the action of a specific regulatory protein. In the absence of a positive regulatory protein, the genes are not expressed.

Primase: An enzyme that is able to initiate the synthesis of a new polynucleotide chain. In DNA replication, primase initiates the synthesis of each Okazaki fragment by producing an RNA primer, which is subsequently elongated by DNA polymerase.

Promoter: A regulatory sequence of DNA at which RNA polymerase binds to initiate transcription.

Prophage: The DNA of a temperate phage that has integrated into a bacterial chromosome via recombination.

Prophase: The first stage of mitosis and meiosis at which the chromosomes condense and become visible. During meiotic prophase I, pairing of homologues and crossing-over occurs.

Pseudolinkage: The apparent linkage between genes located on nonhomologous chromosomes involved in a translocation. Pseudolinkage occurs because the only viable offspring produced by a translocation heterozygote result from alternate segregation gametes, which usually carry the parental combinations of alleles.

Quadrivalent: A group of four homologous or partially homologous chromosomes that are paired with each other during meiotic prophase.

Quaternary Structure: The highest order of protein structure, referring to the structure assumed by an assembly of two or more polypeptide chains to form the functional protein.

Recessive: The classification of an allele whose phenotypic effect is not manifested when heterozygous with a dominant allele.

Recombinant: A progeny cell or offspring, produced as a result of crossing-over, that carries a chromosome with a new combination of alleles for the genes in question.

Recombination: The production of offspring containing combinations of alleles that differ from those found in either parent. Recombination occurs by crossing-over between linked genes.

Recombination Frequency (RF): A measure of the relative distance between genes as determined by the number of times crossing-over occurs between them during meiosis. The RF for a pair of genes is equal to the number of recombinant chromosomes divided by the total number of progeny chromosomes.

Repressed: The state in which the expression of a particular gene or set of genes is prevented by either the effect of a negative regulatory factor or the ineffectiveness of a positive regulatory factor.

Repressible: The condition in which the transcription of a gene or set of genes is prevented by a repressor protein only after it binds to an effector ligand that acts as a co-repressor.

Restriction Enzyme: One of a large collection of enzymes that cleave DNA molecules at specific nucleotide sequences, which differ for each enzyme. Restriction enzymes are isolated from various bacterial species.

Restriction Map: A physical map of a segment of DNA showing the locations of the various restriction enzyme cleavage sites and the distances (in base pairs) between adjacent sites.

RFLPs: Restriction Fragment Length Polymorphisms. Allelic variations in the number or positions of defined restriction enzyme cleavage sites within a population of organisms. RFLPs produce DNA fragments of different lengths when the molecules are cut with specific restriction enzymes.

RNA Polymerase: One of several enzymes that mediate the synthesis of an RNA molecule by catalyzing the formation of phosphodiester bonds between ribonucleotide triphosphates aligned along a DNA template.

Second-Division Segregation: The segregation of a heterozygous allele pair into separate nuclei at the second meiotic division following a crossover between the gene and the centromere during meiosis I. In ordered tetrads, second-division segregation is indicated by a 2:2:2:2 or 2:4:2 spore pattern in which each half of the ascus contains both allelic classes of spores.

Selection Coefficient: A measure of the proportionate reduction in offspring produced by a given genotype relative to that of the most fit genotype in the same population. The fitness of a given genotype is equal to 1 minus the selection coefficient.

Semiconservative Replication: The mode of DNA replication in which each strand of the parental double helix serves as a template for the synthesis of a new strand. The result of semiconservative replication is the production of two daughter molecules, each consisting of one parental and one newly synthesized strand.

Sex Chromosomes: Members of the haploid set of chromosomes (e.g., the X and Y chromosomes in humans) that are involved in the mechanism of sex determination. The content of sex chromosomes and the pattern of inheritance of sex-linked genes are different in the two sexes of a species.

Sex-Linked: The pattern of inheritance of genes located on the sex chromosomes. Sex-linked genes are located on the X chromosome in organisms in which sex is determined by an XX/XY or XX/XO mechanism and on the Z chromosome in organisms in which sex is determined by a ZZ/ZW mechanism.

Spindle Poles: The points at opposite ends of a dividing cell from which the spindle fibers emanate.

Splicing: The mechanism in eukaryotes by which introns are excised from primary transcripts and the remaining exons are joined to produce mature mRNA molecules. The mature mRNAs are then transported to the cytoplasm.

Telocentric: A chromosome in which the centromere is located at the very end.

Telomere: A specialized chromosome region located at the terminal tip of each arm. Telomeres are required to maintain the structural integrity of a chromosome.

Temperate: Refers to phages that do not necessarily lyse a host cell upon infection but can instead integrate their genetic material into the bacterial chromosome to establish a lysogenic relationship.

Template Strand: For a given gene, the strand of DNA that is transcribed into, and therefore complementary to, an RNA molecule.

Tetrad: The set of four chromatids that make up a homologous pair of synapsed chromosomes during meiotic prophase I. Also, the four haploid cells resulting from the meiotic divisions of a single cell.

Tetrad Analysis: The investigation of the meiotic behavior of genes and chromosomes in studies of linkage and crossing-over in fungi by recovery and examination of all four chromatids that participated in a single meiois.

Tetratype: A spore pattern in a fungal tetrad in which four different genotypic classes of spore are found. If the parents are genotypes *AB* and *ab*, a nonparental ditype ascus would contain one *AB*, one *ab*, one *Ab*, and one *aB* spore.

Three-Factor Cross: A cross between one parent that is heterozygous for three linked genes and one that is homozygous for the corresponding recessive alleles. A three-factor cross is designed to determine the linkage relationships among three genes on the basis of the resulting recombination frequencies between them.

Transcription: The process by which a complementary RNA molecule is synthesized upon a DNA template by the action of RNA polymerase.

Transition Mutation: A mutation involving a single base pair substitution in which a purine is replaced by a purine and a pyrimidine is replaced by a pyrimidine.

Translation: The process by which a polypeptide is synthesized upon a ribosome under the direction of the coded instructions provided by an mRNA molecule.

Translocation: A chromosome aberration in which there is an interchange of genetic material between two nonhomologous chromosomes.

Transversion Mutation: A mutation involving a single base pair substitution in which a purine is replaced by a pyrimidine and a pyrimidine is replaced by a purine.

Trihybrid: An individual who is heterozygous for three specified allele pairs.

Trisomic: The condition in a normally diploid cell or organism in which one chromosome is present in three, rather than the normal two, copies. Trisomic cells or individuals have a chromosome content of 2N + 1.

Trisomy: The condition in which one chromosome in an otherwise diploid organism is present in three, rather than two, copies.

Two-Factor Cross: A cross between one parent that is heterozygous for two linked genes and one that is homozygous for the corresponding recessive alleles. A two-factor cross is designed to determine the linkage relationships between two genes on the basis of the resulting recombination frequencies between them.

Uninducible: The inability of a gene or set of genes to be expressed in response to an effector molecule. An uninducible state is caused by a mutation in one of the regulatory components required for transcriptional activation.

Vector: A small, well-defined DNA molecule, such as a plasmid or phage chromosome, used to clone foreign DNA fragments. A vector must be able to replicate itself, and any fragment of foreign DNA attached to it, in an appropriate host cell.

X Chromosome: A sex chromosome in organisms with an XX/XY or XX/XO sex-determination system.

X-Linked: The pattern of inheritance of genes located on the X chromosome of organisms in which sex is determined by an XX/XY or XX/XO mechanism. In such organisms, the sex chromosome content of the two sexes differs, and genes located on the X chromosome have special patterns of inheritance.

Z-Linked: A sex-linked trait in organisms in which females have a ZW chromosome constitution and males have a ZZ chromosome constitution. Examples include birds and butterflies.

ZZ/ZW: The sex chromosome constitution of organisms in which females are the heterogametic sex (ZW) and males are the homogametic sex (ZZ). Examples include birds and butterflies.